Principles of Physical Chemistry

P W Atkins

Fellow and tutor, Lincoln College, Oxford

M J Clugston

Assistant master, Tonbridge School

Pitman

PITMAN PUBLISHING LIMITED
128 Long Acre, London WC2E 9AN

PITMAN PUBLISHING INC
1020 Plain Street, Marshfield, Massachusetts 02050

Associated Companies
Pitman Publishing Pty Ltd, Melbourne
Pitman Publishing New Zealand Ltd, Wellington
Copp Clark Pitman, Toronto

First published in Great Britain 1982

Reprinted 1983

British Library Cataloguing in Publication Data
Atkins, P. W.
    Principles of physical chemistry.
    1. Chemistry, Physical and theoretical
    I. Title II. Clugston, M. J.
    541.3   QD453.2
    ISBN 0–273–01774–8

Printed in Great Britain at The Pitman Press, Bath

Contents

Contents

Preface

We have adopted an approach to physical chemistry which, we believe, makes it easier to understand and more interesting than in conventional treatments. In the first place, we write from the viewpoint of the late twentieth century: we can be confident that the basic features are well established (even though there is still a great deal to understand). In the second place, we think the mathematics, while central to the subject, should be developed hand-in-hand with a physical understanding of what is going on. We always try to impart an insight into atomic and molecular behaviour, and then to use a little mathematics – hardly any is necessary at this level, and certainly none of any subtlety – to turn the insight into something that can be tested experimentally. Finally, we build on a sense of unity. We press home the idea that there is only a handful of basic concepts, and once they have been mastered much of chemistry can be explained simply by exploring their consequences. Physical chemistry should be easy to learn as soon as it is realized that there are only a few, but richly developed, concepts.

A book has to be written in a definite order, but it doesn't have to be used like that. While we have a view on what is the logical order, others will have different ideas, and there may be excellent practical reasons why the material should be presented in a different sequence: individuality is the essence of teaching. So, while we think that the order we use is a sensible one (for reasons we shall develop in a moment) we have tried to write the book so that it can be used readily in almost any order. We have written it so that the chapters fall into short, largely independent, sequences. Within each sequence there is first the presentation and then the development of a concept. This lets the book match a range of ideas on teaching, and minimizes the problem of interweaving the physical chemistry with the other parts of the syllabus.

We begin with something that is almost second nature to most students: the knowledge of the existence of atoms. We look at their properties, arrive at periodicity, and then see how they stick together to form compounds. The chapter on the determination of molecular structure is outside most syllabuses, but we think it appropriate to give a little insight into how, in modern laboratories, molecules are identified and their sizes determined. The next sequence looks at collections of particles, first at gases and then at solids. The third sequence looks at transitions between these states of matter, first of pure materials, and then of mixtures. One special kind of mixture is a solution of ions in water, and we spend a little time on examining their special characteristics, notably electrical conduction. At this point another sequence begins: the study of energy in chemistry. We now move beyond simple physical change to the truly chemical transformation of matter. With the First law of Thermodynamics we begin to see how to use tabulated data to come to useful conclusions about the energy resources of reactions encountered in biology and industry (applications and illustrations from both areas are used throughout the text). The other major role of energy is in the determination of the position of chemical equilibrium, and a sequence of chapters develops this enormously important idea. In the first place we introduce the concept of an equilibrium constant as a quantity that comes from experiment and which characterizes the position of equilibrium. We explore its properties, and see how easy it is to make qualitative predictions about reactions. Then we go on to produce quantitative statements about the actual amounts of substance present in a reaction when it has reached equilibrium.

One entirely optional chapter (in some people's view, because it is outside most syllabuses), but one of central importance to understanding chemistry (in our view), is the chapter on entropy, Chapter 11. Entropy underlies the description of all chemical change, and its meaning is easy to grasp. Once the equilibrium constant has become established as a practically useful idea, it is a simple matter to apply it to a wide variety of systems, and in particular to sort out the processes that lead to the chemical generation of electricity in batteries and fuel cells. The position of equilibrium is independent of the rate at which chemical reactions take place – the confusion between rates and equilibria is something we try to avoid – and so the final chapter looks at this one subject. In this way we travel from topics that are almost pure physics (the structure of atoms) to subjects that are almost biology (the direction of change, and the rate of its occurrence).

We have written the book within severe constraints of length, but we cover the principal existing syllabuses in the UK and elsewhere and the various suggestions that have been put forward for a common core. We have used examination questions from a variety of sources covering (as does the text itself) a broad range of academic ability. We have supplied a summary of each chapter in the hope that if these ideas are remembered the reader will have a framework for answering most questions that are likely to be asked. Since we recognize that science does have a human side, and that readers are often interested both in the way that concepts developed and in the personalities of their discoverers, we have interspersed the text with Boxes containing some historical and biographical details.

We have tried to present a modern book, which is not only pleasant and easy to use, but teaches a great deal about physical chemistry in a lucid and comprehensible way. We could not possibly have done so without the detailed advice of a large number of very experienced teachers. Among them we should like to make special mention of Dr T. P. Borrows, Dr J. N. King, Dr F. J. Marsden, Dr C. L. Mason, Mr P. McManus, Mr D. M. Robins, Mr N. Rowbotham, and Dr J. Spice, who read the manuscript in detail, controlled our wilder impulses, contributed suggestions, and encouraged us to be original where originality was needed. We also wish to thank Professor M. J. Frazer for contributing in invisible but tangible ways. Our thanks are due to Corinne Clugston for reading through the manuscript and to Caron Crisp for typing it. Finally, we should like to thank our publishers for their commitment to the book and for all they have done for its development and production.

P.W.A. M.J.C.
Oxford and Tonbridge, 1981

Acknowledgement

We are grateful to the following boards for permission to use examination questions: The Associated Examining Board, Joint Matriculation Board, Oxford and Cambridge Schools Examination Board, Oxford (Local) Examination Board, Southern Universities Joint Board, The University of London Board, The Welsh Joint Education Committee. We are grateful to the following individuals for permission to reproduce photographs: Professor B. J. Alder, Lawrence Radiation Laboratory, University of California (Fig. 6.1), Professor A. V. Crewe, Enrico Fermi Institute, University of Chicago (Figs 1.1 and 2.1), and Professor M. H. F. Wilkins, Department of Biophysics, King's College, University of London (Fig. 2.2, DNA).

1 Inside atoms

In this chapter we meet the evidence for the modern view of an atom as a central massive nucleus surrounded by a cloud of electrons. We see how the electrons are arranged, and meet the modern description of atomic structure in terms of orbitals. The basis of the periodic classification of the elements is explained in terms of these ideas. We see how atomic spectroscopy is used to identify atoms and explore their structures, and how mass spectrometry is used to determine their masses. Finally we examine the nucleus more closely, and see how it gives rise to the three kinds of radioactivity.

Introduction

The bright dots lying in a line down the photograph in Fig. 1.1 are images of atoms. The idea that matter is built up from these fundamental building blocks is now almost second nature to us, but it is only recently that such direct evidence for their existence has become available. In the late eighteenth and early nineteenth centuries Dalton assembled indirect evidence in favour of the existence of atoms, Box 1.1. Now, in the late twentieth century, we can actually see them.

Dalton thought that atoms were indivisible (which is what *atomos* means in Greek). This view could not be held once electrons were discovered, for if electrons could be knocked out of atoms, atoms had to have an internal structure. Modern chemistry began when it was understood how atoms are built up, because understanding atoms is the key to understanding chemistry.

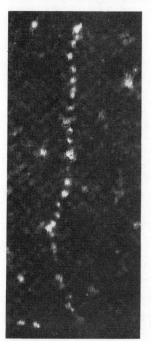

1.1 Electron micro-scope images of atoms

Box 1.1: The origins of the atomic hypothesis

The concept of 'atom' was originated by the Greeks, especially Democritus and Leucippus, and Lucretius's poem 'On the nature of things', which was written in about 56 B.C., uses images that bear a striking resemblance to our modern view. Nevertheless, while it is true that the Greeks put forward the idea that matter is composed of atoms, it was only a speculation. The scientific, as distinct from the philosophical, evidence for the existence of atoms was accumulated about 200 years ago. There were three principal contributions. The *Law of conservation of mass* (expressed by Lavoisier in 1789 just before he fell to the guillotine in 1794) states that matter is neither created nor destroyed in a chemical reaction. The *Law of constant composition* (expressed by Proust in 1799; Proust was the son of an apothecary, and so was surrounded by chemistry throughout his childhood) states that all pure samples of the same compound contain the same elements combined in the same proportion by mass. The *Law of multiple proportions* (Dalton, 1803) states that when the elements A and B combine together to form more than one compound, the various masses of A that combine with a fixed mass of B are in a simple ratio. (Dalton, who worked in Manchester, came to chemistry through his interest in meteorology. He was kindly, socially uncouth, and learned – until he grew tired of reading what others had written.)

The first of these laws suggests that chemical reactions involve the reorganization of things, not their destruction or creation. The second suggests that, since mass characterizes the nature of a compound, the things that are reorganized have characteristic masses. The third suggests that the things are indivisible, so that a compound consists of 1, 2, 3, ... and not some intermediate, non-integral number of them. Dalton accounted for all these observations by proposing the existence of atoms, the indivisible fundamental units of elements.

1.1 The nuclear atom

The first sub-atomic particle to be discovered was the *electron*, Box 1.2. It is negatively charged, and its charge and mass have been measured by observing its deflection in electric and magnetic fields.

When a hydrogen atom loses its single electron a *proton* is left. This is the other fundamental sub-atomic particle of central importance to chemistry. The proton is positively charged (exactly balancing the electron's charge, so that the atom itself is electrically neutral), and its charge and mass can be measured in the same way as for the electron. The proton is about 1840 times heavier that the electron, and so protons (together with the neutrons we meet shortly) account for most of the mass of an atom.

Box 1.2: The discovery of the electron

The discovery of the electron depended on the technological advance of being able to produce high vacua. When that had been achieved in the middle of the 19th century it was not long before *cathode rays* were discovered by passing an electric discharge through a low pressure gas. By 1895 it was known that the cathode rays were negatively charged, but their nature was unknown. The Germans argued that they were waves of some kind; the French and the British thought they were streams of particles. In 1897 J. J. Thomson scored a resounding victory for the latter view by showing that the rays were deflected by electric and magnetic fields, and that the deflection led to a value of charge/mass. Since it had been shown that the rays were absorbed by all materials (by Lenard), Thomson put forward the view that these negatively charged particles, or *electrons* as they became known (the name was coined in 1891 by Stoney), were universal constituents of matter. The charge of the electron was later measured (using Millikan's experiment and the photoelectric effect), and hence the mass itself could be determined. The victorious British–French view survived until 1927, when it was shown that electrons could be diffracted, and hence were waves after all! The crucial experiments were performed by Davisson and Germer and by G. P. Thomson, the son of J. J. Indeed, J. J. Thomson won the Nobel prize for proving that the electron is a particle and his son, G. P. Thomson, received it for proving that the electron is a wave. The resolution of the paradox was achieved with quantum mechanics, which shows how to regard objects, including electrons, as both particles and waves: this is the particle-wave *duality* of matter.

The Geiger-Marsden experiment

In the early years of this century there was a lot of argument about the way protons and electrons were arranged in atoms. The model that originally attracted attention was the 'plum pudding', in which the electrons were pictured as lumps in a jelly-like background of positive charge. This was dismissed by the results of a classic experiment carried out by Geiger and Marsden (in 1909) working under the guidance of Rutherford.

The experiment involved shooting projectiles at atoms and measuring their deflection. As projectiles Geiger and Marsden used α-*particles* (alpha-particles, the nuclei of helium atoms) generated by the radioactive decay of radon. As a target they used a very thin gold foil about 1000 atoms thick. The deflection of the particles by the atoms in the metal foil was detected by observing the bright flashes of light where they struck a fluorescent screen.

They found that most of the α-particles were deflected through angles of less than 1°. Some – about 1 in 20 000 – were deflected through 90° or more, Fig. 1.2. These big deflections eliminated the plum pudding from science. Rutherford expressed his surprise at the result with the remark that "It was almost as incredible as if you fired a 15-inch shell at a piece of tissue paper and it came back and hit you".

Rutherford went on to argue (in 1911) that (*a*) an atom is mostly empty space (to account for the undeflected passage of most particles) but

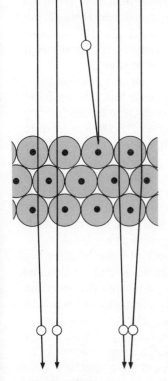

1.2 The Geiger-Marsden experiment

(*b*) most of the mass, and all the positive charge, is in a minute central *nucleus* (to account for the large deflections observed when the α-particle scored an occasional direct hit). The *nuclear atom*, a positive nucleus surrounded by electrons, the former accounting for the atom's mass and the latter for its size, had entered science.

Atomic numbers and atomic masses

The *atomic number* (Z) was originally a label indicating the place of the element in the periodic table. Now it is defined as *the number of protons in the nucleus* of an atom of the element (and so it is also sometimes called the *proton number*). The determination of the value of Z was made possible by Moseley's work on X-rays (in 1913, shortly before he fell in action at Gallipoli). It was known that when elements are bombarded with fast electrons they emit X-rays, and that there are two intense components, the *K and L lines*. Moseley measured their frequencies (ν), and found that when the square root of the frequency is plotted against atomic number, a straight line is obtained, Fig. 1.3. This meant that Z could be found simply by measuring an element's characteristic X-ray frequency and reading off the appropriate value of Z from his graph.

Hydrogen has $Z = 1$ and so it has one proton (and one electron around it). Carbon has $Z = 6$, and its nucleus contains six protons (with six electrons around it). Until 1940 the element with the largest known atomic number was uranium with $Z = 92$, but since then new elements have been made both as a result of nuclear explosions and through the use of particle accelerators, and all the elements up to $Z = 107$ are now known. In general an atom of an element E is written $_Z$E. Therefore a hydrogen atom is written $_1$H and a carbon atom is written $_6$C.

Evidence that the proton is not the only constituent of a nucleus came when the masses of *ionized atoms* (atoms that have lost one or more electrons) were measured using the technique of deflection by electric and magnetic fields. In 1912 Thomson found that neon ($Z = 10$) is a mixture of atoms of different masses. These various alternative versions of an element are called *isotopes* (after the Greek for 'equal place', since they all correspond to the same place in the periodic table).

The explanation of the existence of isotopes became clear when the *neutron* was discovered (by Chadwick in 1932). The neutron is very similar to the proton, having almost the same mass, but it is electrically neutral. Protons and neutrons are collectively known as *nucleons*.

Thomson's observations are explained if a sample of neon contains two types of atom. Both have $Z = 10$ (and hence 10 protons) and are denoted $_{10}$Ne. One type of atom has 10 neutrons in the nucleus and the other has 12. Since there are 10 protons there are also 10 electrons in each case, and so the chemistry of the atoms is the same. The total numbers of nucleons in the two isotopes are 20 and 22, respectively. These are the *mass numbers* (or *nucleon numbers*) of the atoms. The mass number of an element is written A, and an atom of an element of a specified mass number is called a *nuclide* and denoted $_Z^A$E. The isotopes of neon are therefore the two nuclides $_{10}^{20}$Ne and $_{10}^{22}$Ne.

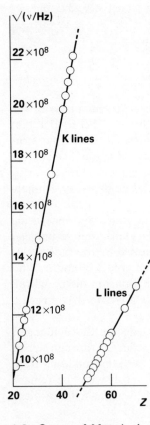

1.3 Some of Moseley's X-ray results

Principles of physical chemistry

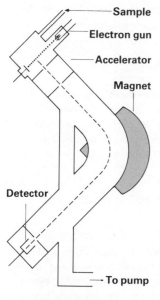

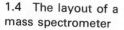

1.4 The layout of a mass spectrometer

The isotopes of an element have different masses because they have different numbers of neutrons. The mass of the nuclide $^{20}_{10}\text{Ne}$ is about 20 times that of $^{1}_{1}\text{H}$, while the mass of $^{22}_{10}\text{Ne}$ is about 22 times greater. This is because a neutron has approximately the same mass as a proton, and so the mass of a nuclide is proportional to its mass number, the total number of protons and neutrons.

Masses and relative masses

The modern method of measuring atomic masses uses the *mass spectrometer*. The principle is illustrated in Fig. 1.4. The sample is introduced in the form of a low pressure vapour (that involves vaporizing a sample if it is liquid or solid). In the ionization chamber it is exposed to an electron beam, which ionizes the atoms by colliding with them and removing one or more electrons. The positive ions so formed are then accelerated through a large potential difference, and the speed they acquire depends on their mass. The beam of fast-moving ions then enters the region of the magnet, where it is deflected, the amount of deflection being greatest for ions of low mass. As the strength of the magnetic field is varied, ions of different mass are focused in turn on to the ion detector. The output from the detector, the *mass spectrum*, is a series of peaks corresponding to the masses present in the sample, the heights giving their relative abundances.

The mass spectrometer actually responds to the mass-to-charge ratio of the ionized species, m/e, and so species of mass m and unit charge contribute to the same peak as those of mass $2m$ and twice the charge. At low ionization beam energies only singly charged ions are produced and so this is not a serious complication.

The mass spectrum of a typical sample of neon is shown in Fig. 1.5. The two peaks corresponding to the isotopes ^{20}Ne and ^{22}Ne are plainly visible, and the abundances 91% and 9% can be deduced from the peak heights. (Modern spectrometers also show a weak ^{21}Ne peak; the abundances are actually 90.92%, 0.257%, and 8.82% for ^{20}Ne, ^{21}Ne, and ^{22}Ne, respectively.) Precise measurements of the positions of the peaks show that the masses of the nuclides are 3.32×10^{-26} kg and 3.65×10^{-26} kg, respectively. The same technique can be applied to other elements, and the masses of all nuclides can be measured.

Instead of working with the actual masses of atoms, and doing calculations on the basis, for example, that the mass of a single atom of ^{12}C is 1.99×10^{-26} kg, it is much more convenient to set up and use a relative scale. The *relative atomic mass*, A_r, is the mass of an atom of a specified nuclide relative to the mass of a single atom of ^{12}C taken as 12.000 0. Therefore the relative atomic mass of the nuclide ^{20}Ne is obtained from

$$\frac{A_r(^{20}\text{Ne})}{A_r(^{12}\text{C})} = \frac{3.32 \times 10^{-26}\ \text{kg}}{1.99 \times 10^{-26}\ \text{kg}} = 1.67.$$

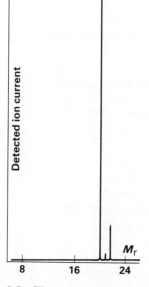

1.5 The mass spectrum of neon

Since $A_r(^{12}\text{C}) = 12.00$, it follows that $A_r(^{20}\text{Ne}) = 20.0$. A typical sample of an element may be a mixture of isotopes, and so its relative atomic mass is an average over all the nuclides present. Average values for typical samples of elements are given in the periodic table printed on the inside back cover. (The *relative molecular mass*, or the *relative formula mass*, M_r, is the sum of the relative atomic masses of the component atoms.)

Example: Predict the form of the mass spectrum of Br_2.

Method: Assume that only singly charged ions are present. There will be Br_2^+ ions arising from the ionization of Br_2 molecules and Br^+ ions from the ionization of Br atoms produced by atomization of Br_2. The naturally occurring bromine isotopes ^{79}Br and ^{81}Br are present in almost equal abundance. Lines can be expected for each isotopic species.

Answer: Species and mass numbers present are

$$^{79}Br^+ \quad ^{81}Br^+ \quad (^{79}Br^{79}Br)^+ \quad (^{79}Br^{81}Br)^+ \quad (^{81}Br^{79}Br)^+ \quad (^{81}Br^{81}Br)^+$$

Mass no.: 79   81   158   160   160   162

The heights of the lines due to ions of mass numbers 79 and 81 will be in the ratio $1:1$. The other four species will also give line intensities in the ratio $1:1:1:1$, because of the equal abundances of the two isotopes, but the lines from $(^{79}Br^{81}Br)^+$ and $(^{81}Br^{79}Br)^+$, both with mass number 160, coincide to give a single line of double intensity. Therefore the diatomic ions give rise to three lines at mass numbers 158, 160, and 162 in the intensity ratio $1:2:1$.

Comment: The positions of the lines can be used to measure the masses of the bromine nuclides. There is no simple way of predicting the relative intensities of different types of ion, and so we cannot predict the intensities of the Br^+ peaks relative to the Br_2^+ peaks.

1.2   The periodic table

Chemists of the nineteenth century had no mass spectrometers, yet they obtained moderately good values of average relative atomic masses (or

Table 1.1   Mendeleyev's predictions for eka-silicon (germanium)

Property	eka-silicon (E)	germanium (Ge)
A_r	72	72.6
Density	5.5 g cm^{-3}	5.3 g cm^{-3}
Melting temperature	high	$937\,°C$
Appearance	dirty-grey	grey-white
Oxide	EO_2; white solid, amphoteric, density 4.7 g cm^{-3}	GeO_2; white solid, amphoteric, density 4.23 g cm^{-3}
Chloride	ECl_4; boiling below $100\,°C$, density 1.9 g cm^{-3}	$GeCl_4$; boils at $84\,°C$, density 1.84 g cm^{-3}

'atomic weights' as they were then called) by measuring the masses of elements that combined together, Box 1.1. A number of people (among them Döbereiner, Newlands, and Meyer) noticed regularities in the properties of elements when they were listed in order of their relative atomic masses. The credit for the first modern form of the periodic table goes, however, to Mendeleyev, who set out the elements in a table and had the confidence to leave gaps for elements then unknown (see Table 1.1), but which seemed to be necessary, Box 1.3. A modern version of the periodic table is on the inside back cover of this book.

Box 1.3: The development of theories about atomic structure

Numerous people had noticed trends and regularities in the properties of the elements, but the credit for organizing them into an arrangement that was powerful enough to be used to make predictions goes to Dmitry Mendeleyev. Mendeleyev was born in Siberia in 1834 and arranged his table in 1869. (According to Russian Orthodox law he was a bigamist, because he had divorced his wife and had married an art student. Tsar Alexander II rejected criticism with the remark that 'Yes, Mendeleyev has two wives, but I have only one Mendeleyev'.) After the discovery that atoms contained electrons, the explanation of chemical periodicity was seen to lie in the way the electrons are arranged. The most important contributions came from Ernest Rutherford. He was a student of J. J. Thomson, having obtained a scholarship to Cambridge from New Zealand (where he had been born in 1871) only because the winner of the competition had dropped out in order to marry. Rutherford arrived in Cambridge when there was high interest in radioactivity, and the study of radioactive rays led to their use as probes of atomic structure. (The direct descendants of this technique are today's particle accelerators, which use fast particles to explore the structure of the fundamental particles: the faster the particle the deeper into the structure of matter it can probe.) Rutherford also achieved the first transmutation of an element, turning nitrogen into oxygen by bombarding it with α-particles. He died in 1937 and lies beside Newton in Westminster Abbey. Rutherford's nuclear atom led Niels Bohr to propose his model of the hydrogen atom. Bohr, a Dane, was working with Rutherford in Manchester at the time. Note the continuity of ideas (and the importance of being in the right place at the right time).

A modern periodic table is constructed on the basis of the atomic numbers of the elements, not their relative atomic masses. That relative atomic mass generally (but not always) increases with atomic number is the coincidence responsible for Mendeleyev's success, but the explanation of periodicity lies in the atomic numbers, for these determine how many electrons an atom possesses. It is to this explanation that we now turn.

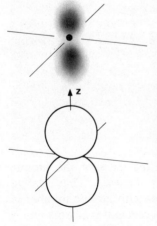

(a)

Normally drawn as

(b)

x

y

z

1.6 (a) An s-orbital, (b) the three p-orbitals

Electrons in atoms

The key to the explanation of chemical periodicity and of all the chemical properties of the elements is the arrangement of electrons around the nucleus. Once Rutherford had established the nuclear atom it was natural for people to speculate that the electrons were arranged like planets around the central nucleus. The speculation was unacceptable, however, because electrons are electrically charged. When anything charged is accelerated (and an orbiting electron, like anything in circular motion, is accelerating all the while) it generates electromagnetic radiation. Therefore a planetary electron would radiate, lose its energy, and spiral into the central nucleus. No electron could survive in orbit for more than a fraction of a second. Consequently, the nuclear atom cannot be a planetary atom.

The first attempt at a modern description of the atom was made by Bohr (1913). In effect he dealt with the problem of electrons collapsing on to the nucleus merely by saying that they couldn't. He asserted that an electron could exist in one of a number of *stationary states*, or *orbits*, and that there was an orbit of lowest energy. An electron in that orbit could not lose energy, and so its spiralling collapse into the nucleus was prevented. Bohr was able to employ the newly emerging *quantum theory* to find a quantitative expression for the energies of the orbits, and they turned out to agree almost exactly with the values obtained from a study of atomic spectra (see below). Nevertheless, his theory did not really explain the structure of atoms, and is now of little more than historical interest.

The modern theory of atomic structure is based on *quantum mechanics*, and in particular on solutions of the *Schrödinger equation*, Box 1.4. The Schrödinger equation is as central to quantum mechanics as Newton's equations are to classical mechanics. The main difference, as far as we are concerned, is that whereas Newton's equations can be solved to predict the precise paths of particles, the solutions of the Schrödinger equation give only *probabilities* of finding particles at various places. So, instead of being able to predict where a particle will be at some instant, we can predict only the chance of it being there.

Box 1.4: The Schrödinger equation

Erwin Schrödinger proposed his equation in 1926 after he had been shown a doctoral thesis by de Broglie (who had proposed a relation between particle properties and wave properties). Schrödinger's first response to the thesis was to say 'That's rubbish'. He was advised to look at it again, realized that it made sense after all, and so set out on the track of his equation. A *Schrödinger equation* is a differential equation of the form

$$-\frac{h^2}{8\pi^2 m}\left(\frac{\mathrm{d}^2\psi}{\mathrm{d}x^2}\right) + V\psi = E\psi.$$

Principles of physical chemistry

h is Planck's constant, *V* is the potential energy, *E* is the total energy, and ψ (the Greek letter psi) is the *wavefunction* which carries all the information about where the particle is located. In atoms wavefunctions are called *orbitals*. When the potential energy has a simple form the equation can be solved exactly: that is the source of our knowledge about the hydrogen atom, where the electron experiences a simple Coulomb potential. In more complicated systems, such as in many-electron atoms and molecules, there is no hope of finding exact solutions, but modern computers are used to obtain highly accurate numerical solutions for the energies and the wavefunctions. Modern theoretical chemistry is largely the exploration of the solutions of the Schrödinger equation for complicated systems. Even the solutions for parts of the DNA molecule can now be obtained.

The structure of the hydrogen atom

The quantum mechanical description of the structure of atoms is best illustrated by the hydrogen atom, which consists of a single electron around a single, massive, central proton. The solutions of the Schrödinger equation for atoms are called *atomic orbitals* (a name selected to convey a less precise impression than 'orbit'). Orbitals give the probability of finding an electron at any point, and in illustrations can be represented by the density of shading. This is the technique used in Fig. 1.6, which depicts some of the orbitals of the hydrogen atom and shows how they are labelled. When the spread of locations of an electron is described by one of these orbitals, we say that an electron *occupies* that orbital. Since the different orbitals represent how the negatively charged electron spreads into different regions of space near the positively charged nucleus, they correspond to different energies. An electron occupying a given orbital therefore possesses a characteristic energy.

When an electron occupies the 1*s-orbital* it has its lowest possible energy. An atom with its electron in this orbital is in its *ground state*. Free hydrogen atoms (such as those responsible for the intergalactic radio waves used in radioastronomy) are almost always found in this state. Figure 1.6 shows that a 1*s-electron* (the electron in a 1s-orbital) is spread spherically symmetrically around the nucleus. The electron should not be thought of as actually circulating round the nucleus but as clustering in its vicinity, with a probability indicated by the density of shading in Fig. 1.6.

The electron can occupy other orbitals, but when it does so it has a higher energy than when it is in the 1s-orbital, Fig. 1.7. The next higher orbitals, the second *shell* of the atom, form a group of four, Fig. 1.6. They fall into two classes: the spherical one is labelled 2s and the three others, each with two lobes, are the 2*p-orbitals*, Fig. 1.6. The lobes can be thought of as lying along *x*, *y*, and *z* axes, and so the 2p-orbitals are labelled $2p_x$, $2p_y$, and $2p_z$, respectively. Higher still in energy is the third shell of the atom; this is a group of nine orbitals (note the progression 1, 4, 9, . . .) all corresponding to the same energy. One is spherical, and is

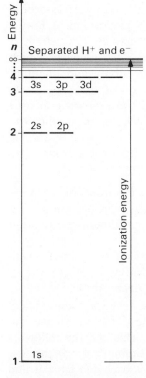

1.7 The energies of the orbitals of the hydrogen atom

called 3s. Three have two lobes like the 2p-orbitals and are labelled 3p. Five have four lobes each and are called the *3d-orbitals*. The s-, p-, and d-orbitals are all we need for most of chemistry, but others with more complicated shapes also occur.

The structure of many-electron atoms

The atomic orbitals of hydrogen are used to describe the structures of more complicated atoms, and we continue to speak of 1s-, 2s-, 2p-, ... orbitals. These orbitals are broadly the same as hydrogen's but differ in detail. For instance, the greater charge of the nucleus draws the inner electrons closer, and so their orbitals spread over a smaller volume. The innermost electron in uranium, for example, is almost 100 times closer to the nucleus on average than in hydrogen, and its orbital is correspondingly more compact. The outer electrons, however, do not experience the full electric charge of the nucleus because it is hidden behind (or *shielded* by) the inner electrons.

There is one important difference between hydrogen and many-electron atoms: the energies of electrons in 2s- and 2p-orbitals are no longer the same, and similarly electrons in 3s-, 3p-, 3d-orbitals have different energies. This is illustrated in Fig. 1.8. The different energies result from the electrostatic repulsions between the electrons.

The electrons of a many-electron atom do not all occupy its 1s-orbital (this is shown by spectroscopy) even though it is the orbital of lowest energy. A remarkable feature of nature is that *no more than two electrons can occupy the same orbital*. This is the *Pauli exclusion principle*, which can be traced to a subtle connection between relativity and quantum theory.

The *building-up principle* (which is also called the *Aufbau principle*) combines these remarks into a set of rules which let us predict the electronic structure of any atom. By 'structure' is meant its *electronic configuration*, the list of the orbitals occupied by its electrons. The principle can be expressed in terms of the following set of rules:

1 If the atom has atomic number Z, then Z electrons have to be accommodated (in the case of the neutral atom; the principle can also be applied to the atom's positive and negative ions by subtracting or adding electrons from Z).
2 Add the electrons one at a time to the lowest energy orbitals available. The order of orbital energies is 1s, 2s, 2p, 3s, 3p, 4s, 3d, 4p.
3 No more than two electrons may occupy any given orbital.

In the case of helium ($Z = 2$) two electrons have to be accommodated. The first enters the 1s-orbital, and the second joins it. Hence the *configuration of helium* in its ground state is $1s^2$. This and the following points are illustrated in Fig. 1.9. In the case of lithium ($Z = 3$) the first two electrons enter the 1s-orbital, but the third is excluded from it and enters the next lowest orbital, which is the 2s-orbital of the second shell; hence the lowest energy *configuration of lithium* is $1s^2, 2s^1$. Beryllium ($Z = 4$) has another electron, and so its configuration is $1s^2, 2s^2$. Boron

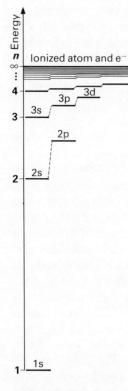

1.8 The energies of orbitals in a typical many-electron atom

Principles of physical chemistry

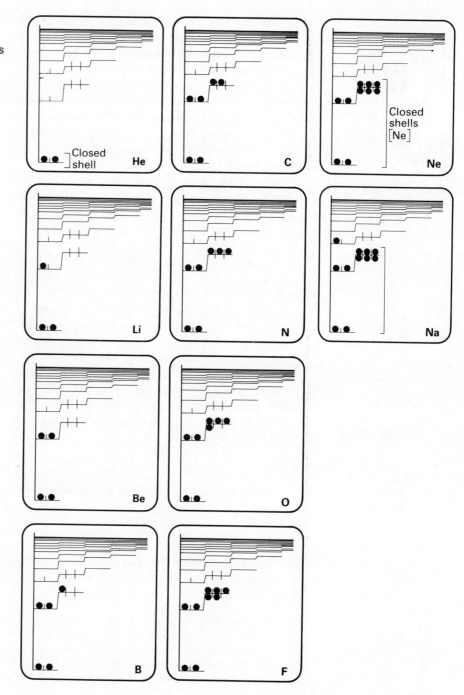

$(Z = 5)$ has one more electron, and so it enters the next lowest orbital, one of the three 2p-orbitals. Its configuration is therefore $1s^2, 2s^2, 2p^1$. There are three 2p-orbitals, and so up to six electrons can enter them. They are therefore gradually filled as we go from boron, carbon, nitrogen, oxygen, and fluorine, to neon. Neon has $Z = 10$, and its configuration is

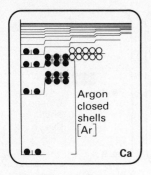

1.10 The configuration of the calcium atom

$1s^2, 2s^2, 2p^6$. This configuration, in which all the orbitals of the shell are full, is called a *closed-shell configuration*. The next element, sodium, has one more electron, which has to go into the next lowest orbital, 3s. Its configuration is therefore $1s^2, 2s^2, 2p^6, 3s^1$.

We have arrived at the explanation of the periodicity of the elements. Lithium $(1s^2, 2s^1)$ has a single s-electron outside an inner, complete shell $(1s^2)$; sodium $(1s^2, 2s^2, 2p^6, 3s^1)$ also has a single s-electron outside a closed shell. The same pattern is repeated between beryllium and magnesium, between boron and aluminium, and so on up to $Z = 18$, argon, which is a closed-shell species, like neon.

The building-up principle also neatly accounts for the transition elements. Potassium $(Z = 19)$ has the configuration $1s^2, 2s^2, 2p^6, 3s^2, 3p^6, 4s^1$, or $[Ar]4s^1$ for short, where $[Ar]$ denotes its *core*, the closed-shell argon configuration. Calcium has the configuration $[Ar]4s^2$, Fig. 1.10. At this point the 3d-orbitals are next in line to be filled. There are five of them, and so they can account for the next ten electrons. Ten is exactly the number of the transition elements of the first long period. At zinc $(Z = 30)$ the 3d-orbitals are full (we say the 3d *sub-shell* is complete), and the next electron enters one 4p-orbital. These 4p-orbitals account for the electrons in the elements up to krypton, which has a closed-shell configuration, $[Ar]4s^2, 3d^{10}, 4p^6$. Krypton resembles argon, and completes this row of the periodic table. All the subsequent periods, including the *inner transition series*, the lanthanides and the actinides (where seven f-orbitals are being occupied with up to 14 electrons), are accounted for in the same way. Since electronic structure is the basis of an element's chemistry, the periodic repetition of configuration accounts for the periodicity of the chemistry of the elements.

Example: State the electronic configurations of the atoms Si and Cl.

Method: Decide how many electrons are involved in each case by noting the atomic numbers (see the inside back cover). Take the preceding closed-shell configuration and write it $[Ne]$ $([Ne] = 1s^2, 2s^2, 2p^6)$; that accounts for 10 electrons. Feed in the remaining electrons, permitting up to 2 to occupy 3s, and up to 6 to occupy 3p. Express the answer in the form $[Ne]3s^a, 3p^b$.

Answer: $Z(Si) = 14$, leaving four electrons to be accommodated outside the closed shell. Two enter 3s, leaving two to enter 3p. Hence Si has the configuration $[Ne]3s^2, 3p^2$. $Z(Cl) = 17$, leaving seven electrons to be accommodated outside the closed shell. Two enter and fill 3s, leaving five to enter 3p. The configuration of Cl is therefore $[Ne]3s^2, 3p^5$.

Comment: Since the outermost partly-filled orbitals of both elements are p-orbitals, both silicon and chlorine are *p-block elements*.

One important property that determines an element's chemical behaviour is the ease with which its outermost, or *valence*, electrons can be removed. The *first ionization energy* is the minimum energy required to remove an electron from a neutral atom in the gas phase (denoted g). That is, it is the energy required for the process

$$E(g) \rightarrow E^+(g) + e^-(g).$$

It is normally expressed as an energy per unit amount (*e.g.*, per mole) of atoms. The *second ionization energy* is the energy required to remove a second electron. It is always larger than the first because more energy is needed to remove an electron from a positively charged ion than from a neutral atom (see Table 1.2).

Table 1.2 First and second ionization energies and electron affinities of atoms

	I_1 kJ mol^{-1}	I_2 kJ mol^{-1}	E_A kJ mol^{-1}
H	1312	—	74
Li	520	7300	56
Na	496	4562	71
C	1086	2353	121
N	1402	2856	−26
O	1314	3388	141
O$^-$	141	1314	−850
S	1000	2258	200
F	1681	3375	333
Cl	1251	2296	349
Br	1140	2084	328

Figure 1.11 shows how the first ionization energy varies for the first few elements and Fig. 1.12 shows the variation throughout the periodic table. There is a pronounced periodicity, the noble gases having high values, showing that the electron is difficult to remove, while the alkali metals have relatively low values, so that relatively little energy is needed to remove the single outermost electron (nevertheless, 500 kJ mol^{-1} is a lot of energy, and so the outer electrons do not simply 'drop off'). Why this is so can be seen from Fig. 1.11, which shows atomic sizes for the first few elements; the atoms shrink across a period (*e.g.*, Li to Ne). This is because the increasing nuclear charge draws in the electrons. As a result they are increasingly difficult to remove. On going from neon to sodium the electron occupies a shell on the outside of the atom and is shielded from the nucleus by the inner electrons. The ionization energy therefore drops back to a smaller value. The small fall in ionization energy between Be and B occurs because the electron being removed is 2s in the case of Be but 2p in the case of B (Fig. 1.9) and the latter is less strongly bound. The drop between N and O is because the O electron shares an orbital with

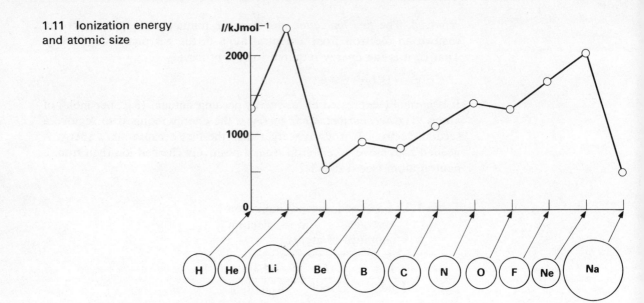

another, Fig. 1.9, which repels it and makes it easier to remove.

One method for measuring the ionization energy of gases involves using a *thyratron*, which is basically a valve containing two electrodes in the gas of interest. The current through the circuit is monitored as the potential difference is increased. When the potential difference is great enough, it strips electrons from the atoms, and the thyratron suddenly becomes highly conducting. Spectroscopy is also used to measure ionization energies, as we describe later.

The *electron affinity*, E_A, of an atom is the change of energy that takes place when an electron is attached to an atom to form a negative ion in the gas phase. That is, it is the energy involved in the process

$$E(g) + e^-(g) \rightarrow E^-(g).$$

The electron affinity is positive when this process is energetically favourable (when energy is released). Halogen atoms have high electron affinities because the added electron enters a gap in a shell and interacts strongly with the nucleus. The noble gas atoms have low electron affinities because the added electron has to enter the orbitals of a new shell a long way from the nucleus and outside the shielding core electrons. See Table 1.2.

1.3 Atomic spectroscopy

The key to understanding the chemistry of the elements is the electronic configurations and energies of their atoms: that is what the periodic table summarizes so neatly. We can check that the atoms have the predicted configurations, as well as explore the energies of electrons, using *atomic spectroscopy*.

The essential feature of spectroscopy is that, when an electron drops from one orbital to another, the discarded energy is carried away as radiation, Fig. 1.13. A ray of light of frequency ν is a stream of *photons*,

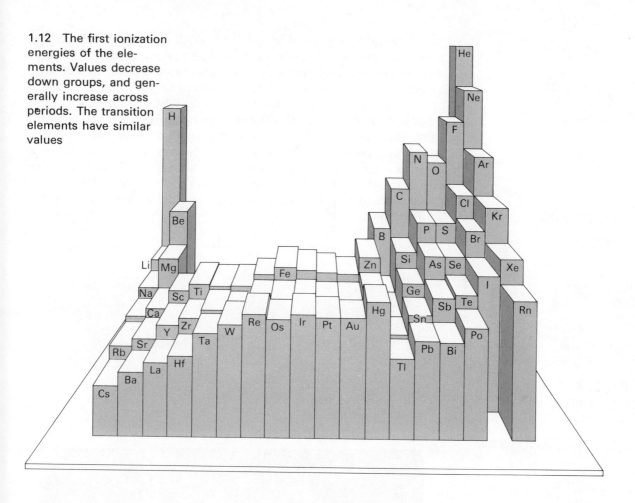

1.12 The first ionization energies of the elements. Values decrease down groups, and generally increase across periods. The transition elements have similar values

each one of energy $h\nu$, where h is *Planck's constant*, a fundamental constant of nature with the value 6.626×10^{-34} J s. An intense ray of light is a dense stream of photons, and a weaker ray contains relatively few. Photons of blue light (which lies at the high frequency end of the visible region of the electromagnetic spectrum) each carry more energy than do photons of red light (which lies at the low frequency end). When an atom loses energy as radiation it does so by generating a single photon of light, and so it is easy to see that the frequency emitted by the atom when it drops from a state of energy E(upper) to one of energy E(lower) is given by the *Bohr frequency condition*

$$h\nu = E(\text{upper}) - E(\text{lower}). \tag{1.3.1}$$

Sometimes it is more convenient to discuss radiation in terms of its wavelength λ (lambda). Wavelength and frequency are related through $\lambda = c/\nu$, where c is the speed of light ($c = 2.998 \times 10^8$ m s^{-1}).

In practice, atoms are generated in their electronically excited states by passing a high-voltage electric discharge through a gaseous sample or by burning a sample in a flame. For instance, the spectrum of atomic

E (upper)

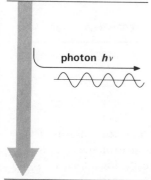

photon *h*ν

E (lower)

1.13 The emission of radiation

hydrogen can be generated in an electric discharge through low pressure molecular hydrogen. The discharge, in effect an electric storm of electrons and ions, rips the molecule apart, and the atoms so formed are initially in a variety of excited states. As their electrons drop to lower states they generate photons of various frequencies, Fig. 1.14, and hence give rise to the observed coloured glow. (The same red glow is seen in sodium street lights when they are first turned on – they contain low pressure hydrogen as well as sodium.) The radiation emitted by the sample is a collection of different colours. If it is passed through a prism or a diffraction grating, Fig. 1.15, the individual frequencies may be resolved and recorded photographically or electronically. That is how the spectrum shown in Fig. 1.16 was obtained.

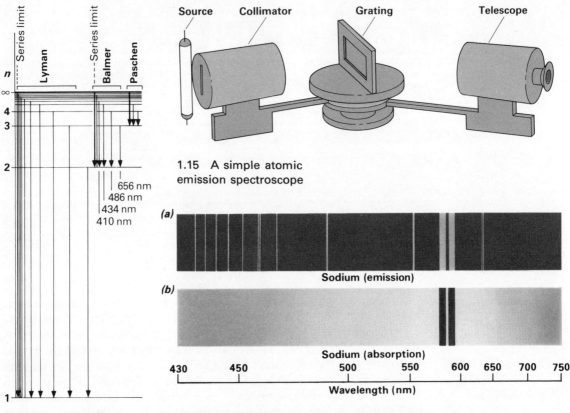

1.15 A simple atomic emission spectroscope

Sodium (emission)

Sodium (absorption)

1.14 The transitions responsible for the spectrum of atomic hydrogen

1.16 The spectrum of atomic sodium in (a) emission, (b) absorption

Atomic spectra can also be observed in *absorption*. If a photon of light of frequency ν hits an atom, it can cause the atom to make a transition from its ground state to an excited state an energy $h\nu$ above it. If the light's frequency ν, and therefore the photon's energy $h\nu$, does not correspond to an excitation energy of the atom, the photon is not absorbed, but it is absorbed if it does match. This means that if the sample is illuminated with a spread of frequencies (*e.g.*, using white light),

Principles of physical chemistry

then there is a reduction of intensity of the beam at every frequency that corresponds to an absorption. If after passing through the sample the beam of light is passed through a prism or a diffraction grating and the spectrum photographed, there will be dark lines, corresponding to the frequencies of the absorbed photons, against an otherwise uniform spectrum of colours. The bright yellow lines in the emission spectrum of sodium atoms shown in Fig. 1.16(*a*) (and which account for the illumination from sodium street lighting once the metal has evaporated) therefore appear as the dark lines in its absorption spectrum, Fig. 1.16(*b*).

Since atoms give rise to characteristic spectra, an obvious application of spectroscopy is to analysis. Helium, for example, was discovered on the sun 30 years before it was identified and isolated on Earth. In the type of spectrometer usually employed in the laboratory a hollow cathode is made of the element of interest, and is heated to high temperature so that it emits its characteristic spectrum. The sample being analysed for the presence of that element is burnt in a cool flame under carefully controlled conditions, and its absorption of the incident light from the hot cathode is monitored electronically with a photomultiplier. A modern commercial instrument can analyse as many as 20 samples for up to 20 elements automatically, and detect concentrations as low as one part in a thousand million (1 in 10^9). The technique is very useful for the detection of undesirable trace elements in water and foodstuffs, and in the analysis of industrial materials. Among the earliest forensic applications of atomic spectroscopy was its use to identify the bullets fired in the St. Valentine's Day massacre in Chicago in 1929.

The analysis of an atom's spectrum lets us find its allowed energy levels and measure its ionization energies. A particularly important case is hydrogen, where late in the nineteenth century Balmer, a Swiss schoolmaster with a passion for numbers, noticed that the spectral lines then known (of wavelengths 656.3 nm, 486.1 nm, 434.0 nm, and 410.2 nm) happened to fit the formula

$$1/\lambda = R_H\{(\tfrac{1}{4}) - (1/n^2)\}, \qquad n = 3, 4, 5, 6 \tag{1.3.2}$$

R_H being a constant. When detectors became available that were sensitive to the infrared and ultraviolet regions of the spectrum, other wavelengths were discovered. All were found to fit a more general form of this expression known as the *Rydberg-Ritz formula*:

$$1/\lambda = R_H\{(1/n_1^2) - (1/n_2^2)\}. \tag{1.3.3}$$

R_H is the *Rydberg constant* for hydrogen (its value is $1.097 \times 10^7 \, \text{m}^{-1}$). Different values of n_1, which is called a *quantum number* (more precisely, the *principal quantum number*), correspond to different series of lines in the spectrum. For instance, there is the *Lyman series* ($n_1 = 1$ and $n_2 = 2$, 3, 4, ...) with wavelengths in the ultraviolet region, the *Balmer series* ($n_1 = 2$ and $n_2 = 3$, 4, 5, ...) with wavelengths in the visible, and the *Paschen, Brackett, Pfund, and Humphreys series* ($n_1 = 3$, 4, 5, and 6, respectively, with n_2 starting with the value $n_1 + 1$ in each case) which lie in the infrared.

The Rydberg-Ritz formula can be deduced from the Schrödinger equation, and the value of the Rydberg constant can be predicted very accurately. The agreement between theory and experiment for both hydrogen and more complex atoms shows beyond reasonable doubt that our ideas on atomic spectra and structure are now substantially correct (but we still have a long way to go in understanding how to apply them!).

If we refer to Fig. 1.14 we see that the wavelengths of the Balmer series come closer together the further along the energy series we look. The same is true of the Lyman series. The line of shortest wavelength of each series is called the *series limit*. In an absorption experiment using wavelengths shorter than the series limit every photon carries so much energy that when it is absorbed it ionizes the atom. This is the basis of the spectroscopic method of measuring ionization energies. On comparing Fig. 1.14 with Fig. 1.7 it is clear that the Lyman series limit corresponds to the ionization energy of the atom. Therefore, all we have to do to find the ionization energy is to note the wavelength of the Lyman series limit and convert it to an energy.

1.4 The nucleus in chemistry

Most of chemistry takes place without any change occurring in the structure of the nucleus; it plays out its role by governing the tightness with which the electrons are bound. Nevertheless nuclei can change their states. Sometimes they simply change their energy and, like atoms, emit the excess as radiation. Sometimes they eject charged particles, and as a consequence change their own charge, and therefore their identity. These changes and the corresponding emissions are called *radioactivity*.

Radioactivity: α-, β-, and γ-rays

The early investigators of radioactivity soon found that radioactive emissions could be classified into three groups. They were found to respond differently to electric and magnetic fields, Fig. 1.17.

β-*rays* (or β-*particles*) were easily identified because they behaved in exactly the same way as electrons, which were already known. They are fast electrons ejected from within the nucleus. Since they carry unit negative charge, they leave the nucleus with one additional unit of positive charge. Therefore, when an element emits a β-particle it changes from atomic number Z to atomic number $Z+1$, Fig. 1.18. For instance, the nuclide $^{14}_{6}C$ is β-active (it is a *radionuclide*), and every atom that loses an electron from its nucleus becomes an atom of $_7N$. Note that, because the electron is so light, the mass number does not change on β-emission; ^{14}C therefore decays into ^{14}N, the common isotope of nitrogen. In general an element $^A_Z E$ *transmutes* into $_{Z+1}^{A}E'$ on β-emission.

γ-*rays* are undeflected by either electric or magnetic fields, they are *electromagnetic waves*, and a part of the electromagnetic spectrum. They are of much higher frequency than visible light, and hence of much shorter wavelength. Their wavelengths are shorter even than the wavelengths of X-rays (shorter even than 100 pm). They are very penetrating, and damage living tissue by causing ionization. One industrial

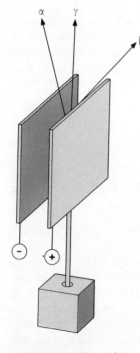

1.17 The effect of an electrostatic field on α-, β-, and γ-radiation

Principles of physical chemistry

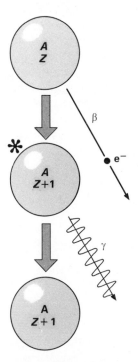

1.18 β-decay (accompanied by γ-emission)

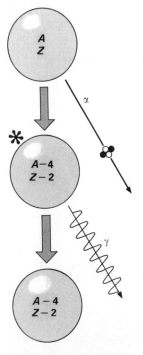

1.19 α-decay (accompanied by γ-emission)

application is to the measurement of thickness, and to the detection of flaws in metal objects.

A photon of γ-radiation is generated when a nucleus changes its energy state and the excess energy is carried away as radiation. Since the nucleons are bound together by the *strong force* (as distinct from the weaker electrostatic force between charged particles that accounts for the electronic structure of atoms), the energies involved, and therefore the frequencies, are much greater than in atomic spectra. γ-rays usually accompany β-rays (and α-rays) because the latter's emission leads to the formation of a new element in an energetically excited state, Fig. 1.18. Since a γ-ray is the result of an energetic state of a nucleus losing energy, its generation does not result in the formation of a new element.

α-rays (or *α-particles*) were the most difficult to identify because nothing similar was known when they were first discovered. For one thing, they were deflected by an electric field in a way that showed they were positively charged. The crucial experiment was inspired, once again, by Rutherford. It showed that when radon decayed by α-emission, helium was formed in the vicinity. Since the mass-to-charge ratio of α-particles had been measured in a deflection experiment, the conclusion could be reached at once that an α-particle is $_2^4\text{He}^{2+}$, or the bare helium nucleus, a tight cluster of two protons and two neutrons. That combination of nucleons is very stable, just like a closed-shell electronic structure. In some atoms there is evidence that protons and neutrons group together into α-particles inside the nucleus.

When an α-particle is emitted by a nucleus it carries away two units of positive charge. As a result the atomic number of the element drops from Z to $Z-2$, Fig. 1.19. Since the α-particle consists of four nucleons it carries away four mass units, and so the mass number of the element also changes from A to $A-4$. An element $_Z^A\text{E}$ therefore transmutes into $_{Z-2}^{A-4}\text{E}''$ on α-emission.

Example: Predict what nuclides are produced on (*a*) β-decay of $_{11}^{24}\text{Na}$ and (*b*) α-decay of $_{88}^{226}\text{Ra}$.

Method: Use the rules stated above: on β-decay $_Z^A\text{E}$ changes to $_{Z+1}^A\text{E}'$ and on α-decay $_Z^A\text{E}$ changes to $_{Z-2}^{A-4}\text{E}''$. Identify the element from the information in the table on the inside back cover.

Answer: (*a*) $A=24$, $Z=11$; therefore the product is $_{12}^{24}\text{E}'$. The element with $Z=12$ is Mg. Hence the decay is $_{11}^{24}\text{Na} \rightarrow {}_{12}^{24}\text{Mg} + \beta$.

(*b*) $A=226$, $Z=88$; therefore the product is $_{86}^{222}\text{E}''$. $Z=86$ identifies the product as Rn. Therefore the decay is $_{88}^{226}\text{Ra} \rightarrow {}_{86}^{222}\text{Rn} + \alpha$.

Comment: The radium decay reaction was used by Rutherford and Royds to generate the radon in their identification of the α-particle.

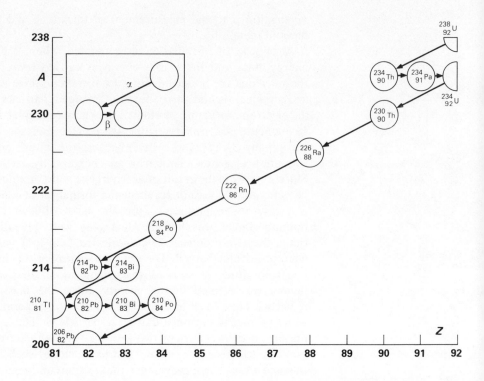

The transmutations of the elements that occur when the various types
of radiation are emitted by unstable nuclei are illustrated very well by the
uranium series, Fig. 1.20. Other natural radioactive elements of mass
numbers similar to uranium's decay down to some isotope of lead, $_{82}$Pb,
and lead's unusual stability has led people to speculate that element 114
would also be stable, if only it could be prepared. Elements are syn-
thesized in the interiors of stars and during their explosions. On Earth we
have to use less dramatic methods, and new nuclides are prepared by
collisions in particle accelerators. The *nuclear reaction* that takes place on
impact is written in the form

$$\begin{array}{c} \text{target} \\ \text{nuclide} \end{array} \left(\begin{array}{cc} \text{incoming} & \text{outgoing} \\ \text{particle,} & \text{particles} \end{array} \right) \begin{array}{c} \text{product} \\ \text{nuclide.} \end{array}$$

The 'equation' is balanced so that the sums of the atomic numbers and
the mass numbers are the same before and after the collision; electrons,
protons, and neutrons are written $_{-1}^{0}\text{e}$, $_{1}^{1}\text{p}$, and $_{0}^{1}\text{n}$.

Example: State the nuclide produced in the following nuclear reactions:
(a) $_{7}^{14}\text{N}(_{0}^{1}\text{n}, _{1}^{1}\text{p})$; (b) $_{4}^{9}\text{Be}(_{2}^{4}\alpha, _{0}^{1}\text{n})$; (c) $_{3}^{7}\text{Li}(_{1}^{1}\text{p}, _{2}^{4}\alpha)$.

Method: The sum of the mass numbers in the products must equal their
sum in the initial species; the same is true of the atomic numbers. Therefore
balance the reaction by finding $_{Z}^{A}\text{E}$ for which this is true, and identify the
nuclide from the inside back cover.

Principles of physical chemistry

The energy released in a nuclear reaction is the basis of the generation of nuclear power. In *nuclear fission* collision with a neutron results in the fracture of a heavy nucleus with the liberation of nuclear binding energy (and a consequent overall loss of mass through $E = mc^2$) and even more neutrons. These neutrons in turn cause fission in several neighbouring nuclei, and a *chain reaction* sets in. If a lot of the neutrons are absorbed (as in a reactor) the chain reaction is controlled. If they are not absorbed the chain reaction rapidly escalates into an explosion. In *nuclear fusion* energy is released when light nuclei fuse into a single nucleus, as in the interiors of stars and soon, it is hoped, under controlled conditions on earth.

Radioactive half-lives

The time it takes for half the radioactive atoms in any sample to decay is called the radioactive *half-life* of the radionuclide and is denoted $t_{\frac{1}{2}}$. Half-lives can vary from less than 1 s (*e.g.*, 3.04×10^{-7} s for $^{212}_{84}Po$) to more than a thousand million million years (*e.g.*, 6×10^{15} years for $^{50}_{23}V$).

The *law of radioactive decay* is that *the rate at which a radioactive sample decays is proportional to the number of radioactive atoms present:*

$$\text{rate} = \lambda N. \tag{1.4.1}$$

λ (lambda) is called the *decay constant* and is related to the half-life by $\lambda = (\ln 2)/t_{\frac{1}{2}}$. Why this should be the law will become clear once we have dealt with reaction rates (in Chapter 14): essentially it means that nuclei fall apart at random. The importance of the law is that it enables us to predict the proportion of atoms of some initial sample that have not yet decayed. This follows from the solution of the last equation for the number of atoms present at time t given that N_0 were present initially:

$$N = N_0 e^{-\lambda t}. \tag{1.4.2}$$

The form of this *exponential decay* is shown in Fig. 1.21.

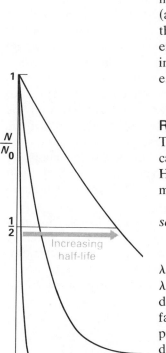

1.21 Radioactive decay

Method: Use eqn (1.4.2) to find the fraction N/N_0 remaining at $t = 20$ yr. For λ use $(\ln 2)/t_{\frac{1}{2}}$.

Answer: $\lambda = (\ln 2)/(5.25 \text{ yr}) = 0.132 \text{ yr}^{-1}$. Then from eqn (1.4.2),

$$N/N_0 = e^{-(0.132 \text{ yr}^{-1}) \times (20 \text{ yr})} = e^{-2.64} = 0.0714, \text{ or } 7.14\%.$$

Comment: This gives an impression of the great lengths of time it takes for some dangerous radionuclides to disappear. Check that the fraction of ^{60}Co does not fall below 0.01% until 70 years after the explosion.

A well-known and important application of the last result is in the dating of archeological remains. The nuclide ^{14}C is continuously generated in the atmosphere as a result of the bombardment of nitrogen by cosmic rays. Therefore, to a good first approximation, its abundance is constant, and the fraction in CO_2 is also constant (about 1 molecule in 10 000). Since living things continuously ingest carbon compounds as complex molecules or as CO_2 (in photosynthesis) they contain a constant fraction of ^{14}C. When they die they stop ingesting, and so the ^{14}C content is no longer replenished but decays according to the radioactive decay law with a half-life of 5570 years. Therefore, if the proportion of ^{14}C to ^{12}C is determined (by measuring the β-activity of the object) at some date after the object's death, the lifetime can be determined. The error in the determination arises partly from the fact that the proportion of ^{14}C in the atmosphere is not quite constant because it depends on things like solar sunspot activity and the earth's magnetic field. That is why it is necessary to calibrate the method against a more direct dating method, such as counting tree rings. Other nuclides (such as ^{40}K) are also used in archeological dating procedures.

There are numerous advantageous applications of radioactive isotopes in chemistry (as well as in medicine). In particular radioactive *labelling* lets us follow the course of an element through a chain of reactions. This is especially important in biochemistry, where metabolic pathways can be unravelled by following the course of ^{14}C from its ingestion to its excretion.

Summary

1 The atom consists of a central nucleus surrounded by electrons. This *nuclear atom* was suggested by the *Geiger-Marsden α-particle scattering experiment.*

2 The identity of an element is determined by its *atomic number, Z,* the number of protons in its nucleus. Different *isotopes* of an element have different numbers of neutrons but the same number of protons.

3 The *mass number* of an atom is the total number of *nucleons* (protons and neutrons) in the nucleus. A *nuclide* is an atom of specified mass number.

4   The *relative atomic mass*, A_r, of an element is the mass of one atom of a specified nuclide relative to the mass of one atom of ^{12}C taken as 12.

5   Mass numbers, actual masses, and abundances of nuclides are most readily determined using the *mass spectrometer*.

6   *Atomic orbitals* give the distributions of electrons in atoms.

7   *s-orbitals* are spherically symmetrical, *p-orbitals* are double-lobed, and *d-orbitals* are four-lobed.

8   The *electronic structure* of atoms is expressed in terms of the *configuration* predicted using the *building-up principle*. This is based on the *Pauli exclusion principle*, that no more than two electrons may occupy an orbital.

9   *Chemical periodicity* arises from the periodicity of the electronic configurations of the elements, and that periodicity comes naturally from the building-up principle.

10   *Ionization energies* are periodic, and are related to the electrostatic control of the nucleus over its outermost, *valence*, electrons.

11   The configurations $1s^2$ and $1s^2, 2s^2, 2p^6$ are examples of *closed-shell* configurations. *Closed-shell* species are stable mainly because they are so compact and all the electrons are close to the nucleus.

12   *Atomic spectroscopy* is used to fingerprint elements, to investigate the electronic configurations and energy levels of atoms, and to measure ionization energies. The analysis is based on the *Bohr frequency condition*.

13   Light of frequency ν and wavelength $\lambda = c/\nu$ may be regarded as a stream of *photons*, each one of energy $h\nu$, where h is *Planck's constant*.

14   The spectrum of atomic hydrogen is summarized by the *Rydberg-Ritz formula*, eqn (1.3.3).

15   Radioactive emissions from nuclei fall into three groups. *α-rays* are bare helium nuclei, $^4_2He^{2+}$. *β-rays* are electrons ejected from the nucleus. *γ-rays* are electromagnetic radiation, generally of wavelengths less than about 100 pm.

16   When an α-particle is emitted a nuclide $^A_Z E$ *transmutes* to $^{A-4}_{Z-2}E'$; when a β-particle is emitted a nuclide changes from $^A_Z E$ to $^A_{Z+1}E'$; when a γ-ray is emitted a nucleus is changing its energy but not its identity.

17   The *radioactive half-life* is the time it takes for the abundance of a radionuclide in the sample to decrease by half. The *law of radioactive decay* implies an exponential decrease in the abundance of a radioactive species.

Problems

1   Write an essay on the development of the modern theory of atomic structure.

2   State the numbers of protons, neutrons, and electrons in each of the following atoms: (*a*) ^{16}O, (*b*) ^{18}O, (*c*) ^{27}Al, (*d*) ^{35}Cl, (*e*) ^{40}Ca, (*f*) ^{235}U.

3   Write an essay on isotopes. Include an account of their discovery, their significance, the measurement of their relative atomic masses, and their uses.

4   On the basis that a sample of sulphur had the following composition, calculate its average relative atomic mass: ^{32}S, $A_r = 31.972$, 95.0%; ^{33}S, $A_r = 32.972$, 0.8%; ^{34}S, $A_r = 33.969$, 4.2%.

5   Describe the appearance of the mass spectrum of argon.

6 Give the electronic configurations of the ground states of the following atoms: (a) O, (b) Si, (c) K, (d) Fe.

7 A 100 W light bulb gives out about 2×10^{18} photons of yellow (550 nm) light each second. What is the energy carried by each photon? What fraction of the total energy emitted by the lamp do they carry? (Use $1\,W = 1\,J\,s^{-1}$.)

8 Explain why the minimum energy to remove an electron from the ground state of a hydrogen atom can be calculated from the Rydberg-Ritz formula by setting $n_1 = 1$ and $n_2 = \infty$. Calculate the molar ionization energy of hydrogen.

9 The longest wavelength radiation that can be used to ionize a sodium atom in its ground state is 241 nm. What is the molar ionization energy of sodium?

10 What nuclides are formed by the following radioactive decays: (a) β-decay of ^{131}I, (b) α-decay of ^{251}Cf, (c) β-decay of ^{90}Sr?

11 A ^{60}Co source was installed in a hospital radiation unit in 1980. Given that the half-life of the nuclide is 5.25 years, calculate the fraction of the source that will remain in (a) 1990, (b) 2080.

12 (a) The relative atomic mass of the element palladium is 106.40. State concisely what this statement means.

(b) Using mass spectrometry, the element gallium has been found to consist of 60.4% of an isotope of atomic mass 68.93 and 39.6% of an isotope of atomic mass 70.92. Calculate, to three significant figures, the relative atomic mass of gallium. *(Oxford)*

13 (a) Explain the meaning of the term *first ionization energy* of an atom.

(b) The first, second, third, and fourth ionization energies of the elements X, Y, Z are given below:

Ionization energies in $kJ\,mol^{-1}$

	First	Second	Third	Fourth
X	738	1450	7730	10550
Y	800	2427	3658	25024
Z	495	4563	6912	9540

Using this information, state, giving your reasons, which element is most likely
  (i) to form an ionic univalent chloride;
  (ii) to form a covalent chloride;
  (iii) to have +2 as its common oxidation state.

(Oxford)

14 (a) Describe briefly with the aid of a sketch the spectrum of atomic hydrogen.

(b) Explain how this spectrum provides evidence for
  (i) electron energy levels in the atom,
  (ii) the ionization energy of hydrogen.

(AEB 1979 part question)

15 (a) List the three main fundamental particles which are constituents of atoms, and give their relative masses and charges.

(b) Similarly, name and *differentiate* between the radiations emitted by naturally occurring radioactive elements.

(c) Complete the following equations using your periodic table to identify the elements X, Y, Z, Q, and R. Add atomic and mass numbers where these are missing:

$(i)\ ^{24}_{11}Na \rightarrow X + ^{0}_{-1}e$ $(ii)\ ^{14}_{7}N + ^{1}_{0}n \rightarrow ^{14}Y + _1Z$
$(iii)\ Si \rightarrow ^{27}_{13}Q + ^{0}_{+1}e$ $(iv)\ R + ^{4}_{2}He \rightarrow ^{13}_{7}N + ^{1}_{0}n$.

(d) Refer to c(i) and c(iii) above. For each of these two processes, briefly describe ONE chemical test which could be used to confirm that a change of chemical element has occurred.

(*SUJB*; *continued as qu.* 10.20)

16 (*a*) Explain what is meant by each of the following terms:
   (*i*) electron, (*ii*) proton, (*iii*) neutron, (*iv*) isotopes.
   (*b*) The atomic number provides three pieces of information about an element. What are they?
   (*c*) The radioactive atom $^{224}_{88}$Ra decays by α-emission with a half-life of 3.64 days.
       (*i*) What is meant by 'half-life' of 3.64 days?
       (*ii*) Referring to the product of the decay, what will be its mass number and its atomic number?
       (*iii*) Radium is in Group II of the periodic table. In what Group will the decay product be?
   (*d*) Explain briefly the principles underlying
       (*i*) the use of radioactive isotopes as 'tracers'
       (*ii*) the dating of dead organic matter using radiocarbon, $^{14}_{6}$C.

(*London*)

17 Discuss the evidence for the existence of isotopes. The nuclear reaction

$$^{6}Li + {}^{2}H \rightarrow 2\,^{4}He$$

is proposed as a possible future energy source. Given the isotopic masses ^{6}Li (6.015 06), ^{2}H (2.014 07) and ^{4}He (4.002 63) calculate the energy released by consumption of 1 g ^{2}H. What daily consumption of ^{2}H corresponds to a power of 1 MW?

Outline a scheme for the production and concentration of ^{2}H, explaining the relevant physical and chemical principles.

(*Oxford Entrance*)

18 Discuss the atomic spectrum of hydrogen and its relation to our understanding of the electronic structure of atoms.

Suggest explanations for the following observations:
   (*a*) The atomic spectrum of hydrogen contains lines in the radiofrequency region of the electromagnetic spectrum.
   (*b*) A line in the spectrum of atomic hydrogen on a distant object in the universe occurs at a wavelength of 300 nm though it is known to occur in the laboratory at 121.6 nm.

(*Oxford Entrance*)

2 Inside compounds

In this chapter we see why atoms stick together and form compounds. We see that ionic bonds and covalent bonds can be explained in terms of electrostatic interactions between electrons and nuclei. There are various ways of describing how electrons take part in bonding, and we shall see how to explain the formation of molecules. We see why molecules have characteristic shapes, and meet some of the special properties of double bonds, aromatic molecules, and metals.

Introduction

The world around us takes on its richness because atoms cluster together into compounds. The view that molecules are definite arrangements of atoms grew during the nineteenth century as a result of careful observations and measurements in the laboratory. Modern techniques have brought about a revolution in the nature of the evidence. Originally molecules were suggested in order to explain observations on the masses of elements that combined together, and in order to account for observations on chemical reactions and the wide range of organic materials, Box 2.1. Now we can see them. An example of the kind of resolution currently attainable with an electron microscope is shown in Fig. 2.1. The coiled strand of atoms in the molecule is just visible.

2.1 Electron microscope image of haemoglobin

Box 2.1: The development of the concept of molecule

The name molecule was introduced by Avogadro as a diminutive of the Latin *moles*, mass. The line symbol for a chemical bond, as in H—Cl, was first used in 1858, and in the paper that introduced it the author, A. S. Couper, introduced what we now call a 'structural formula'. The first reference to a 'molecular structure' seems to be in 1861 by the Russian chemist Butlerov. Ball-and-stick models were used in 1865, and so by then (by the time that Kekulé was proposing his structure for benzene) the idea of molecules as having not only a definite composition but also a definite architecture was well established.

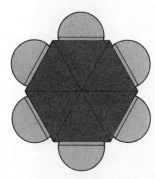

Once the structures of atoms were understood, attempts were made to explain molecular structure and bond formation in their terms. A principal contribution was made by G. N. Lewis (b. 1875) when he identified the importance of the electron pair. Much credit for the rationalization of bonding theory in terms of quantum mechanics goes to Linus Pauling (b. 1901, who was rejected as a Rhodes Scholar, but went on to win Nobel prizes for both chemistry and peace) who exerted great influence on chemical thought through his book, 'The nature of the chemical bond'. Modern studies of the chemical bond depend on careful experimental determinations of structure, using spectroscopy or X-ray diffraction, and detailed theoretical studies using computers. Now we can compute the details of the electron distributions in complex molecules and predict their bond angles and bond lengths. The major outstanding problem is the prediction of their reactivities, but even that is beginning to become possible. Even though we have such computing power, the problem of interpreting, understanding, and using the results remains as challenging as ever.

The main questions are why compounds form, and why molecules have such characteristic shapes. This field of chemistry is called *valence*. We have to explain why some atoms, those of the noble gases, are only rarely found in compounds while others, particularly carbon, readily form networks of chains and rings. We must also explain why some molecules are straight (like CO_2) while others are bent (like H_2O). Some atoms do not appear to form single, well-defined molecules, but stick together to form huge arrays. Common salt is one example, diamond another.

An impression of the range of molecular structures is given by the drawings of molecular models in Fig. 2.2. The atoms have been drawn approximately to scale, and so the illustration gives an idea of the range of sizes of molecules.

The smallest molecule shown is that of hydrogen. Since it consists of two identical atoms it is called a *homonuclear diatomic molecule*. The oxygen molecule is also a homonuclear diatomic molecule. It is significantly bigger than molecular hydrogen because there are more electrons surrounding the two nuclei. The hydrogen chloride molecule is an example of a *heteronuclear diatomic molecule*; note the great difference in the sizes of the two atoms. Water, ammonia, and methane are simple examples of *polyatomic molecules*. The benzene molecule is one of the most important molecules in organic chemistry because it is the parent of the aromatic compounds, species having an unexpectedly high stability. Virtually the extreme of molecular complexity is shown by the structure of the genetic material DNA, and the illustration shows only a fragment of this huge molecule. The complexity and size of DNA are related to its function: it has to convey detailed coded structural information from one generation to the next.

2.2 Molecular models of H_2, O_2, HCl, H_2O, NH_3, CH_4, C_6H_6, and (overleaf) DNA

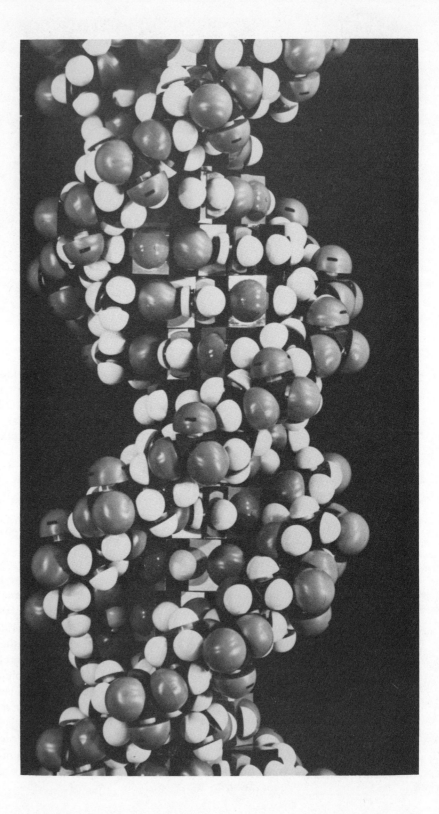

2.2 (*continued*)
A small section
of DNA

Principles of physical chemistry

2.1 General features of bond formation

Atoms stick together and form compounds if by so doing energy is released. The energies of atoms and their compounds are determined largely by the way their electrons are arranged, and so compound formation may occur if a rearrangement of the electrons leads to the attainment of a state of lower energy.

The positions of electrons can change in one of two ways. An atom may transfer one or more of its electrons completely to the other. Alternatively, two atoms may share electrons. When complete transfer occurs the atoms A and B become the *ions* A^+ and B^- (or A^{2+}, B^{2-}, *etc.* if more than one electron is transferred) and then stick together as a result of the electrostatic attraction between opposite charges. The link so formed is an *ionic bond* (or an *electrovalent bond*). When atoms form a bond by sharing electrons the link is called a *covalent bond*. We shall look at both types of bond in turn, explain when to expect one or the other, and show, in this chapter and in Chapter 5, how the properties of compounds depend on the nature of their bonds.

The nature of the ionic bond

A and B stick together as a result of the formation of an ionic bond if the unit A^+B^- lies lower in energy. (We concentrate on the simplest case first, when only one electron is transferred.)

There are three stages in ionic bond formation. Each one involves energy, and only if the *overall* energy change is favourable will the bond form. The first step is the removal of an electron from A to form the positive ion A^+, Fig. 2.3. This involves investing the ionization energy of the atom A (Section 1.2). The second step is the addition of the electron to B to form the negative ion B^-. This involves the electron affinity of B (Section 1.2). If the electron affinity is positive (as for Group VII atoms) this step helps to lower the total energy. The final step is the lowering of energy as a result of the electrostatic attraction between the oppositely charged ions. It follows that an ionic bond is most likely to form when A has a low ionization energy (so that it is energetically easy to remove an electron), when B has a high electron affinity (so that some energy is recovered when the electron attaches to B), and when the ions are close together (so that the Coulomb interaction is strongly favourable).

Atoms with relatively low ionization energies are those of Groups I and II. They can lose their valence electrons with relatively little energy investment, and so can be expected to take part in ionic bonding. Once they have lost their valence electrons the energy involved in removing any of the core electrons is too great to be recovered and so core electrons do not take part in bonding.

The atoms with relatively high electron affinities are those of Group VII because the added electron enters a gap in the valence shell and can interact favourably with the opposite charge of the nucleus. A closed-shell species such as a halide ion or a noble gas atom has a relatively low (or even a negative) electron affinity because the additional electron has

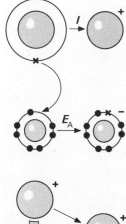

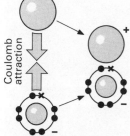

Coulomb attraction

2.3 The formation of an ionic bond

to enter a new shell confined to the outside of the atom, where it can interact only weakly with the distant, shielded nucleus.

This discussion suggests that an ionic bond is likely to form between the atoms on the left and the atoms on the right of the periodic table, and that electron transfer occurs until the atoms involved have lost or gained enough electrons to reach closed shells. A summary of this behaviour is that *atoms appear to have a tendency to reach a noble gas configuration.* This is the basis of the *octet rule* formulated by the early chemists (especially G. N. Lewis in 1916) and expressed by them in terms of atoms tending to acquire a *stable octet* of electrons (the s^2, p^6 closed shell of modern atomic theory).

The octet rule should not be taken to mean that Group I and II atoms 'want' to lose electrons, or that Group VI and VII atoms 'want' to gain them. It means that the electrostatic interaction between ions is strong enough to overcome the combined effects of the ionization energy and the electron affinity so long as only the valence electrons are involved. Going beyond closed shells – removing core electrons and attaching electrons to closed shells – involves so much energy that it cannot be recovered from the interaction between the ions that would result, and so *overall* there is no energy advantage.

Dot-and-cross diagrams summarize ionic bond formation in a simple way. They can be used to discuss all kinds of ionic species, such as A^+B^-, $A^{2+}(B^-)_2$, and so on. The valence electrons are shown as dots for one atom and as crosses for the other. Then the gaps in the valence shell of B are filled by transferring valence electrons from A. This is illustrated in the following *Example.*

Example: Show how the electrons are arranged in the following compounds: (*a*) KBr, (*b*) K$_2$O, (*c*) CaCl$_2$, (*d*) MgO.

Method: As all the species are compounds of metals from Groups I or II with non-metals from Groups VI or VII, they are ionic in character. Account for their bonding in terms of the complete transfer of electrons: transfer a sufficient number to give closed shells on both the positive and negative ions. Note that Group I elements can lose one electron, Group II can lose two; Group VI can accept two electrons, and Group VII can accept one. Clarify the origin of the elctrons by using **x** to denote the electrons from one atom and ● to denote the electrons from the other. Show only the outermost electrons.

Answer: (*a*) $\left\{ \text{K}\times \quad \bullet\ddot{\text{Br}}\colon \right\} \rightarrow \left\{ \text{K}^+ \quad (\times\ddot{\text{Br}}\colon)^- \right\}$, or K^+Br^-

(*b*) $\left\{ \text{K}\times \quad \bullet\ddot{\text{O}}\bullet \quad \times\text{K} \right\} \rightarrow \left\{ \text{K}^+ \quad (\times\ddot{\text{O}}\times)^{2-} \quad \text{K}^+ \right\}$, or $(\text{K}^+)_2\text{O}^{2-}$

(*c*) $\left\{ \colon\ddot{\text{Cl}}\bullet \quad \times\text{Ca}\times \quad \bullet\ddot{\text{Cl}}\colon \right\} \rightarrow \left\{ (\colon\ddot{\text{Cl}}\times)^- \text{Ca}^{2+} (\times\ddot{\text{Cl}}\colon)^- \right\}$, or $\text{Ca}^{2+}(\text{Cl}^-)_2$

(*d*) $\left\{ \times\text{Mg}\times \quad \bullet\ddot{\text{O}}\bullet \right\} \rightarrow \left\{ \text{Mg}^{2+} \quad (\times\ddot{\text{O}}\times)^{2-} \right\}$, or $\text{Mg}^{2+}\text{O}^{2-}$.

Principles of physical chemistry

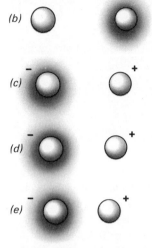

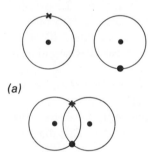

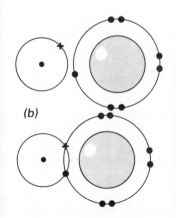

2.4 Electron density changes during the formation of an ionic bond

2.5 The formation of a covalent bond in (a) H_2, (b) HCl

Although dot-and-cross diagrams are easy to draw and simple to use they are little more than an accounting method. Nevertheless, they do capture the principal features of ionic bond formation, the transfer of electrons and the electrostatic attraction between ions. That such a transfer does take place is confirmed experimentally by X-ray diffraction of ionic solids (Chapter 5) and supported theoretically by solving the Schrödinger equation for a pair of atoms as they approach, Fig. 2.4. This type of computation shows vividly that initially (at a) the two atoms have their usual numbers of electrons; both are neutral (and Li is bigger than F because the fluorine nucleus, $Z = 9$, has a strong control over its electrons). At (c) the three energy contributions listed above favour electron transfer, and the electron hops on to the F, making it slightly larger, and leaving the small helium-like Li^+ ion nearby. As the Li^+ ion moves closer it begins to distort the electron distribution of the F^-, which it tends to draw into its vicinity. At this stage (e) the Li^+ ion is beginning to recover a share in its lost electron, and so the bond is starting to lose its purely ionic character and to take on the characteristics of a covalent bond. This emphasizes the fact that *no bond is purely ionic*.

The nature of the covalent bond

Covalent bonds are formed when two atoms share pairs of electrons. The dot-and-cross description of covalent bonds is shown in Fig. 2.5. Each atom contributes one electron to the bond, but instead of complete transfer (as in an ionic bond) the atoms share the pair. In this way both atoms reach a closed-shell configuration, but without the expenditure of energy needed to transfer an electron completely from one atom to the other.

Example: Show how the electrons are arranged in the following compounds: (a) H_2O; (b) CO_2; (c) HCl.

Method: All the compounds consist of covalent molecules, and hence the bonding electrons are shared. Draw diagrams showing shared electrons, the number of bonds being such as to complete the valence shell of each atom involved. Clarify the origin of the electrons by using x for the electrons from one atom and • for the electrons from the other.

Answer: (a) H_2O has two covalent single bonds

$$Hx \quad \bullet \ddot{O} \bullet \quad xH \rightarrow H\overset{\bullet\bullet}{\underset{\bullet\bullet}{\times O \times}}H \quad \text{or} \quad H \diagdown O \diagup H$$

2.6 Electron density distribution in benzene

The experimental confirmation that electron pairs are shared comes from a variety of sources. The most striking (and the one that gives a clue to the reason why the pair holds the atoms together) is X-ray diffraction. We describe the basis of this method in the next chapter; all we need now is one of the pictures it gives. Figure 2.6 shows the output from an X-ray diffraction study of benzene. The density of shading represents the density of electrons. The familiar hexagonal shape of the molecule is clearly visible (compare Fig. 2.2). The six carbon atoms can be picked out as regions of high electron density, but what is important for the present discussion is that *between* the carbon atoms there are also regions of high electron density. The contour lines indicates that electrons are likely to be found there, which is just what we expect on the basis that neighbouring atoms share pairs of electrons.

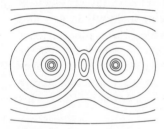

2.7 Electron density distribution in H_2

Benzene has a lot of electrons, and some peculiar properties, and so we put it aside for the moment and return to it later. The simplest molecule of all is the hydrogen molecule, H_2, which contains only two electrons. An electron density plot for hydrogen is shown in Fig. 2.7. This diagram also shows an electron-rich region between the two nuclei, just as in benzene, and so once again the simple H $\overset{\bullet}{\times}$ H picture of the covalent bond is confirmed. In this case, though, we can begin to see why the molecule is stable. Since the electron pair lies between the two positively charged nuclei it attracts both, and acts as a kind of 'electrostatic glue'. That is the basic reason for the strength of covalent bonds: when the electron pairs are shared they lie between the nuclei, attract them both, and bind them together.

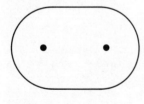

H —————— H

2.8 Three ways of drawing a covalent bond

There are several ways of drawing a bond between two atoms. The most detailed description involves plotting the electron density, as in Fig. 2.7. That is much too detailed for most applications, and so usually only the general shape of the electron density distribution is indicated. This is shown in Fig. 2.8. The sausage shape (which is called a *molecular orbital*) indicates the region where the electron pair is most likely to be found (but the line should not be thought of as suggesting a sharp boundary). Even this representation of a bond is too detailed for most purposes, and so it is common to use a straight line to denote the presence of an electron pair. In this simplest representation a hydrogen molecule is written as H—H. Always bear in mind, however, that a line between two atoms is only

Principles of physical chemistry

shorthand and the *explanation* of the strength of a bond has to be found in the details of the electron distribution of the type shown in Fig. 2.7.

The structures of polyatomic molecules

Polyatomic molecules are covalent molecules built from three or more atoms, and include H_2O, CH_4, and DNA. They can all be discussed in terms of the sharing of electron pairs. The main problem is to account for their shapes. Why, for instance, is H_2O bent, CO_2 straight, and CH_4 tetrahedral?

The shapes of molecules are a result of the electrostatic repulsions between electron pairs. The four electron pairs of the methane molecule repel each other, and so move to positions where they are furthest apart, which is at the corners of a regular tetrahedron. Therefore the CH_4 molecule is expected to be tetrahedral, as is found experimentally. The most favourable arrangements of other numbers of electron pairs are summarized in Fig. 2.9. We shall make use of them later.

We saw in the case of the hydrogen molecule that there were various ways of representing the bond, ranging from the full electron density plot to the shorthand of a simple line. The same types of drawing may be used in the case of methane. The most detailed representation is shown in Fig. 2.10(*b*), showing the electron density in the molecule. The

2

3

4

5

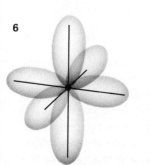

6

2.9 The energetically most favourable ways of arranging pairs of electrons

(a)

H ₓ• ×H
  •C•
H ×•× H

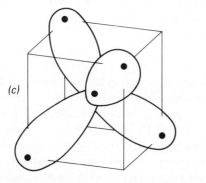

(c)

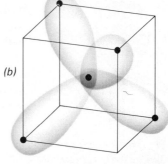

(b)

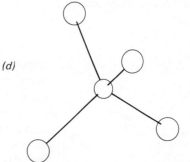

(d)

2.10 Four ways of drawing the structure of CH_4

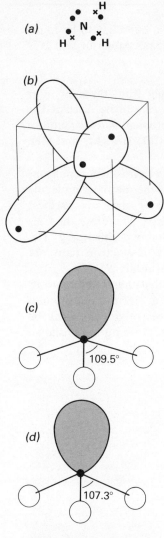

(a)

(b)

(c) 109.5°

(d) 107.3°

2.11 Three ways of drawing the structure of NH₃ and (at the bottom) the experimental bond angle

2.12 The electron pair distribution in NH₃

densest regions are at the atoms, and the dark regions between the carbon atom and each of the four protons correspond to the electron pairs. The general shape of the electron distribution can be represented by the four balloon-shaped regions shown in Fig. 2.10(c). The simplest representation is in terms of a line diagram, Fig. 2.10(d), where each line indicates where the electron pair is most likely to be found.

The shape of the ammonia molecule, NH_3, can be predicted using the same kind of argument. The five electrons in the valence shell of the nitrogen atom can complete their octet by sharing a further three electrons from three hydrogen atoms. This is shown in a dot-and-cross diagram in Fig. 2.11(a). As in methane there are four electron pairs, and so under the effect of electrostatic repulsions they move to the corners of a tetrahedron. Three of the electron pairs have protons attached, the fourth is a *lone pair*. At this stage we predict that ammonia should be a regular tetrahedron, with one of the corners occupied by a lone pair of electrons: this shape is called *pyramidal*.

Experiments on ammonia show that the prediction is nearly but not quite right. The molecule is pyramidal, but the H—N—H angles are 107.3°, not the 109.5° of a regular tetrahedron, Fig. 2.11(d). (This example shows that while the physical basis of chemistry might be well established, we can still trip up when we apply the principles to real problems.)

The extra piece of information we need is that lone pairs of electrons have a greater repulsive effect than bonding pairs. This can be seen on the basis of the electron density plot for ammonia in Fig. 2.12. It shows that the lone pair of electrons spreads over a greater volume than the bonding pairs (because the latter are pinned down by the protons). Because they spread through a greater volume they drive other electron pairs away more effectively. The modern version of the *valence-shell electron pair repulsion* (or VSEPR) description of molecular structure develops this point. It accounts for the shapes of molecules on the basis that the repulsive effects of electron pairs are in the order:

lone pair/lone pair
stronger than lone pair/bond pair
stronger than bond pair/bond pair.

The shape of the NH_3 molecule can now be explained. As a first approximation it is a regular tetrahedron. That prediction is based on equal repulsions between all four pairs. In fact the effect of the lone pair on the bond pairs is strongest, and so it forces the three bond pairs to move away, Fig. 2.13. The three N—H bonds therefore move closer, like an umbrella closing, and the angle between them is reduced below 109.5°, as observed. Another point about ammonia is that a proton may attach readily to the lone pair and form the ammonium ion, NH_4^+. All four N—H bonds are now the same, and so the ion is predicted to be tetrahedral. This is confirmed by experiment.

The water molecule, H_2O, is the next species to consider. The dot-and-

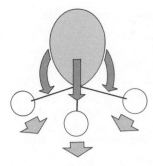

2.13 The effect of electron pair repulsions in NH_3

cross diagram is shown in Fig. 2.14(*a*). There are four electron pairs in the valence shell, and so, as a first approximation, the molecule is predicted to be tetrahedral, two of the corners being occupied by protons, and two by two lone pairs.

Now consider the effect of the different strengths of the repulsions. The two lone pairs repel each other and so tend to move apart, Fig. 2.15. There is a kind of scissor action hinged on the oxygen; as the lone pairs move apart the bonds move closer together. Since the lone pair–lone pair repulsions are so strong, they dominate the bond pair repulsions and force the O—H bond pairs closer together. In other words, the bond angle is predicted to be less than the tetrahedral value of 109.5°, in accord with the observed bond angle of 104.5°.

Lone pair–lone pair interactions also account very neatly for the weakness of the F—F bond in F_2. Each fluorine atom has three lone pairs of electrons, and the remaining electron pair forms a bond. Because of the strong lone pair interactions between the two atoms we can predict that the bond will be very weak compared, for example, with the bond in H_2, where there are no lone pairs. The values of the *dissociation energies* (the energy required to break the bond), 158 kJ mol^{-1} for F_2 and 432 kJ mol^{-1} for H_2, support this conclusion.

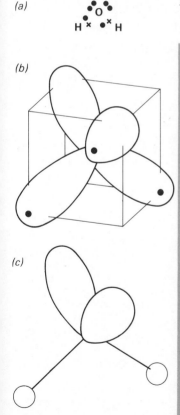

(a)

(b)

(c)

2.14 Three ways of drawing the structure of H_2O

Example: Predict the shapes of the following molecules: (*a*) CCl_4; (*b*) H_2S; (*c*) SF_6; (*d*) PH_3.

Method: Consider only the valence electrons. Base the answer on the repulsions between lone pairs of electrons and bonding pairs. Decide how many bonding pairs there are on the central atom; then decide how many lone pairs are centred on it. Refer to Fig. 2.9, decide on the most symmetrical shape for the species, and then allow for VSEPR repulsions.

Answer: (*a*) Carbon has 4 valence electrons, each one involved in a bond. Hence there are 4 bonding pairs in CCl_4 and no lone pairs (except on the chlorines, which are irrelevant to the discussion). From Fig. 2.9 we see that a regular tetrahedral arrangement is the likely structure, with a Cl atom at each point of the tetrahedron and the carbon atom in the centre.
(*b*) Sulphur has 6 valence electrons. In H_2S there are two bonding pairs, and so there are two lone pairs (accounting for the remaining 4 electrons). The basic shape is therefore tetrahedral, two of the points of the tetrahedron being occupied by H atoms, and the other two points being occupied by lone pairs of electrons. The lone pairs exert the dominant repulsive effects, and so the shape will distort and resemble that of H_2O.
(*c*) Sulphur has 6 valence electrons. There are 6 bonds in SF_6, and so all six electrons are involved in covalent bonding and there are 6 bonding pairs of electrons. Figure 2.9 shows that the shape of the molecule is a regular octahedron, with an F atom at each point.
(*d*) Phosphorus has 5 valence electrons. In PH_3 three of the electrons are involved in bonds, and so there are 3 bonding pairs. The remaining two electrons form a lone pair. Three bonding pairs and one lone pair give rise to a tetrahedral shape, with three atoms at three corners of the tetrahedron,

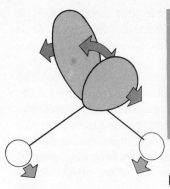

and the fourth position occupied by the lone pair. The lone pair dominates, and the actual shape will resemble that of NH_3, i.e., pyramidal.

Comment: The pungency of sulphur and phosphorus containing compounds (such as the bad eggs smell of H_2S) is due to the presence of the lone pairs of electrons. The mechanical repulsion between closely growing walnuts results in bunches like those predicted for electron pairs by VSEPR.

2.15 The effect of electron pair repulsions in H_2O

Multiple bonds

Carbon forms single, double, and triple bonds, and this versatility accounts for the richness of the world of organic compounds. Just as a single bond is formed when two atoms share an electron pair, a *double bond* is formed when they share two pairs, Fig. 2.16(*a*), and a *triple bond* is formed when they share three pairs, Fig. 2.16(*b*).

The arrangement in space of the electron pairs in multiple bonds can be worked out on the basis of the electron repulsions. In a double bond the electron pairs arrange themselves tetrahedrally around each atom (as a first approximation) and so the electron density distribution in a molecule such as ethene is as shown in Fig. 2.17(*a*). This is normally simplified to the line diagram shown in Fig. 2.17(*b*).

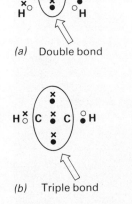

(a) Double bond

(b) Triple bond

2.16 (*a*) A double bond (*b*) a triple bond

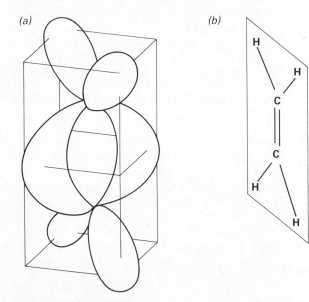

(a)   *(b)*

2.17 (*a*) Electron distribution in ethene (*b*) the conventional bonds

Two properties of the double bond can be understood on the basis of Fig. 2.17(*a*). In the first place the electron pairs of a double bond are not

Principles of physical chemistry

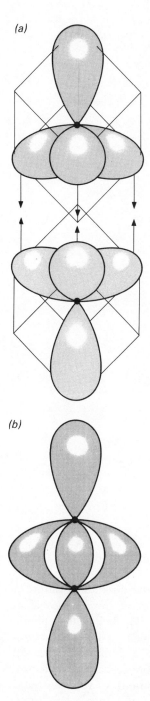

(a)

(b)

2.18 (a) The formation of a triple bond (b) the resulting electron density distribution

in the best place to bring about bonding. In a single bond they are directly between the atoms; in a double bond they are off-centre. Therefore a double bond is expected to be weaker than two single bonds. This is observed, and the dissociation energy of a double bond between two carbon atoms is $612\,kJ\,mol^{-1}$, while twice a single bond's dissociation energy is $696\,kJ\,mol^{-1}$. A double bond is therefore predicted (and found) to be reactive because it is energetically favourable to break it open to leave one C—C single bond and to form two new single bonds with the reactant. The other property is that a double bond resists twisting: it is *torsionally rigid.* This is because in order to twist one CH_2 group relative to the other (in ethene) the four electron pairs on one atom would have to be forced to lie in the same plane: that brings them closer together and increases their repulsions.

A carbon–carbon triple bond can be pictured as in Fig. 2.18. The electron pairs of the atoms are arranged at the points of a tetrahedron, and the three pairs involved in the triple bond are at the three shared corners. Although this arrangement of electrons can stick the two atoms together, each bond is even less favourable than the arrangement in a double bond. Triple bonds are therefore very reactive, and their dissociation energy ($837\,kJ\,mol^{-1}$) is much less than that of three single bonds ($1044\,kJ\,mol^{-1}$). Note that the arrangement of electrons in two triply-bonded carbons, Fig. 2.18, predicts that a molecule X—C≡C—Y should be *linear.* This is found in practice, and ethyne, H—C≡C—H, has all its atoms in a straight line.

Aromatic stability

The dot-and-cross diagram for benzene is easy to construct, and is shown in Fig. 2.19. It predicts an alternation of single and double bonds round the ring, and so we should expect to see an alternation of electron density (and an alternation of bond lengths) in the electron density plot. Reference back to Fig. 2.6 shows that this is not the case, for every carbon–carbon region has the same electron density. Furthermore, in organic chemistry we soon meet the fact that benzene behaves quite unlike what would be expected on the basis that it contained three double bonds.

The reason for benzene's stability can only be partly understood in terms of dot-and-cross diagrams, and the true explanation has to be found in terms of quantum mechanics. This shows the limitation of the diagrams. The partial explanation is as follows. First, consider Fig. 2.19. Why did we arrange the electron pairs as shown there, and not as in Fig. 2.20? The electron pairs are not in fact confined to particular bonds, but are mobile, and can spread uniformly around the ring, Fig. 2.21. This has the effect of reducing the electron repulsions within the molecule, and so lowers its energy. The spreading also allows the bond lengths to settle into a regular planar hexagonal arrangement, which also reduces the repulsions. The spreading of electrons round the ring is called *electron delocalization,* and the extra stability is called *delocalization stabilization* (and sometimes *resonance stabilization*).

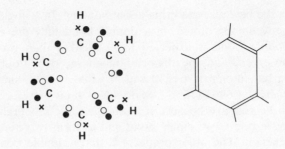

2.19   One structure for C_6H_6

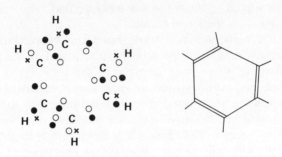

2.20   Another structure for C_6H_6

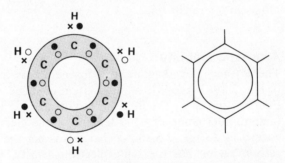

2.21   The delocalized bond structure of benzene

2.2   Other points about bonding

The outstanding feature of covalent bonding is the directional nature of the bonds. There are specific locations for the atoms where they share electron pairs most effectively. As a result, the molecule has its minimum energy when the atoms are in precisely determined positions; therefore we can speak of covalent molecules as having a definite shape, with definite bond lengths and bond angles. Furthermore, once all the electron pairs have been shared there is no further possibility of forming bonds with other molecules. As a consequence, covalently bonded compounds tend to be well-defined, discrete (but not necessarily small) clusters of atoms. Typical examples are the molecules of water, the hydrocarbons, proteins and DNA. Some crystals, such as diamond, can be regarded as colossal single covalently bonded molecules.

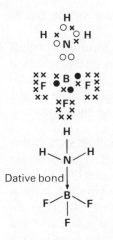

2.22 The formation of a dative (or coordinate) bond

In contrast, ionic bonds do not have definite directional characteristics because the electrostatic Coulomb force is the same in every direction. Furthermore, a single ion may attract more than one ion of opposite charge, and so extensive ionic aggregates may form. That is why ionic materials, like common salt, are generally found as extensive crystalline arrays.

Dative bonds

A *dative bond* is formed when one atom (A) contributes both electrons to a bond. An example is the bond between ammonia and boron trifluoride, BF_3, leading to the formation of the *addition compound* $H_3N \rightarrow BF_3$, Fig. 2.22. There is no fundamental difference between this type of bond and the normal covalent bond, and the basis of its stability is exactly the same. The only difference is the source of the electrons used in building the bond.

Polar bonds and electronegativities

Covalent bonds may be classified as *polar* or *nonpolar*. In a nonpolar bond there is no net redistribution of charge, and the atoms are electrically neutral. The bonds in molecular hydrogen and in all homonuclear diatomic molecules are nonpolar; each C—C bond in benzene is non-polar. A polar bond arises when the formation of a bond leaves partial charges on the atoms. An ionic bond may in fact be regarded as an extreme form of a polar covalent bond, because an electron is wholly transferred from one atom to another, resulting in the formation of A^+B^-. Less extreme are the normal polar bonds, when electrons tend to drift slightly from one atom to another, leaving one with a small positive charge and the other with a small negative charge. The hydrogen chloride molecule, for instance, has a polar covalent bond, there being a small shift of electrons towards the chlorine to give $H^{\delta+}$—$Cl^{\delta-}$; there is a similar shift in water, resulting in the charge distribution $_{\delta+}H \diagdown ^{O^{2\delta-}} \diagup H^{\delta+}$; the six C—H bonds in benzene are all slightly polar. The presence of polarity in a bond indicates

Table 2.1 Pauling electronegativities, χ

H						
2.20						
Li	Be	B	C	N	O	F
0.98	1.57	2.04	2.55	3.04	3.44	3.98
Na	Mg	Al	Si	P	S	Cl
0.93	1.31	1.61	1.90	2.19	2.58	3.16
K	Ca					Br
0.82	1.00					2.96
Rb	Sr					I
0.82	0.95					2.66

that it should not be thought of as being strictly purely covalent, but as having a touch of ionic character.

A convenient measure of the tendency of an atom to form polar bonds is its *electronegativity*, the relative power of an atom to attract a shared pair of electrons. This is normally denoted χ (chi) and its value is proportional to the mean of the ionization energy and the electron affinity of the atom. A list of values is given in Table 2.1 and their variation across the periodic table is illustrated in Fig. 2.23.

An atom with a high ionization energy (which makes its existing electrons difficult to remove) and a high electron affinity (which means that it is energetically favourable for other electrons to attach to it) has a high electronegativity. When such an atom takes part in bonding, it tends to acquire electrons and not to lose them, and so it tends to form the negative end of a polar bond; in an extreme case it forms a negative ion. Typical examples are the halogen atoms. In contrast, an atom with a low electron affinity (a low energy advantage on acquiring electrons) and a low ionization energy (energetically easy to lose electrons) has a low electronegativity. Such an atom tends to lose its valence electrons to a more electronegative atom, and therefore to become the positive end of a polar bond, and may even become a positive ion. Typical examples are the alkali metals, see Table 2.1.

The electronegativities of atoms allow us to predict the degree of polarity of bonds. Electrons tend to accumulate near the atom with the greater electronegativity; therefore such atoms tend to acquire a small negative charge. The direction of the bond polarity can therefore be

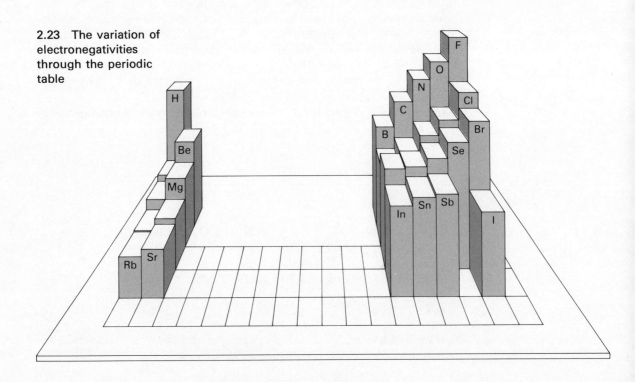

2.23 The variation of electronegativities through the periodic table

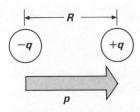

2.24 The structure of an electric dipole of magnitude $p = qR$

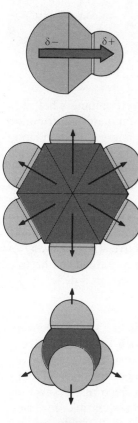

2.25 HCl and H_2O are polar, C_6H_6 and CH_4 are nonpolar molecules

predicted very simply by comparing electronegativities. For instance, the electronegativities of H and C are $\chi(H) = 2.20$ and $\chi(C) = 2.55$, which indicates a drift of charge towards the carbon atom in a C—H bond, which should therefore be thought of as being of the form $C^{\delta-}$—$H^{\delta+}$. The electronegativity difference is greater in the case of an O—H bond, and although the sense of polarization is still $O^{\delta-}$—$H^{\delta+}$ there is a greater drift towards the oxygen. This accounts for the pronounced polarity of the water molecule, a polarity that is of extreme importance for the role that water plays as a solvent.

Dipole moments

A measure of the polarity of a molecule is provided by the quantity known as its *dipole moment, p*. The definition of a dipole moment is illustrated in Fig. 2.24: if a positive charge q and a negative charge $-q$ are separated by a distance R, then the magnitude of the dipole moment is $p = qR$. If an isolated NaCl unit is regarded as a pair of ions with charges $+e$ and $-e$ (where $-e$ is the charge of an electron) then its dipole moment is of magnitude eR, R being the bond length. It is known (from spectroscopy) that $R = 251$ pm, and so we can predict that

$$p \approx (1.602 \times 10^{-19}\,\text{C}) \times (251 \times 10^{-12}\,\text{m}) = 4.02 \times 10^{-29}\,\text{C m}.$$

Dipole moments are normally expressed in *debye*, D, a unit named after a pioneer in the study of polar molecules: 1 D corresponds to 3.336×10^{-30} C m. In the present case the dipole moment is 12.1 D. This is huge compared with that of most molecules because the NaCl unit is ionic. The hydrogen chloride molecule, for instance, has $p = 1.08$ D.

The dipole moment of H_2 is zero, because the bond is nonpolar. The dipole moment of an O—H bond arises from its polarity, and has the value 1.3 D. This is much smaller than the dipole moment of the NaCl unit because O—H is only $O^{\delta-}$—$H^{\delta+}$ rather than O^-H^+. The dipole moment of a C—H bond is even smaller, about 0.5 D, because it is less polar still.

The overall dipole moment of a molecule may be zero even though it contains polar bonds. This is because the dipole moments of each bond may be arranged so symmetrically that they cancel. The six dipole moments of the C—H bonds in benzene, Fig. 2.25, are arranged symmetrically, and opposite pairs cancel leaving the molecule nonpolar overall. The four C—H dipole moments of methane are arranged symmetrically in three dimensions, and although it is harder to see that cancellation occurs, it does in fact, and overall the molecule is also nonpolar. The water molecule, on the other hand, being bent, has two O—H bond dipole moments that do not cancel, and as a result it is polar ($p = 1.84$ D).

Example: Which of the following molecules have dipole moments: (*a*) HCl; (*b*) Cl_2; (*c*) CO_2; (*d*) $CHCl_3$?

Method: Base your answer on the symmetry of the molecule. Homonuclear diatomic molecules and symmetrical linear molecules cannot have electric dipole moments. In the case of polyatomic molecules decide whether the shape is such that the dipole moments of the bonds cancel.

Answer: (a) HCl is a heteronuclear diatomic molecule, and therefore has a dipole moment. Since Cl is more electronegative than H, the polarity of the molecule is $H^{\delta+}Cl^{\delta-}$.
(b) Cl_2 is a homonuclear diatomic molecule and therefore has no dipole moment.
(c) CO_2 is a linear, symmetrical polyatomic molecule, and therefore has no dipole moment. Alternatively, answer on the basis that the symmetry of the molecule is such that the dipole moments of the two C=O bonds cancel.
(d) $CHCl_3$ is tetrahedral, but the C—H dipole moment does not cancel the three C—Cl dipole moments. It is therefore polar. Since the Cl are more electronegative than the H, the H atom is the positive end of the dipole, and the Cl atoms are the negative ends.

Comment: The experimental values of the dipole moments of the molecules are (a) 1.08 D, (b) 0, (c) 0, and (d) 1.01 D.

The dipole moment is of great importance in determining whether a species will act as a solvent for ions. This is because a major contribution to the dissolution of an ionic compound is the energy of interaction between the solvent molecules and the ions. In a *polar liquid* (a liquid composed of polar molecules) such as water, the negative end of the molecular dipole (the O atom) is strongly attracted to positive ions, and the positive end, the pair of H atoms, is strongly attracted to negative ions. A molecule like benzene, with no net dipole moment, cannot interact in the same way. Consequently while benzene, a *nonpolar liquid*, cannot dissolve ionic materials, water is very effective. (A quick way of deciding whether or not a liquid is polar is to bring a charged rod up to a stream of it running from a burette; if it is polar the stream is deflected.)

The metallic bond
A block of metal is an ordered array of positive ions in a sea of electrons. From the point of view of bonding theory, a block of metal is like a hydrogen molecule, but on a bigger scale. Conversely, a hydrogen molecule is like a tiny fragment of a metal: it is two positive ions (H^+) in a sea of electrons, but the sea, consisting of only two electrons, is now only a puddle. The mobility of the sea in metals is responsible for their characteristic conductivity, reflectivity, ductility, and malleability, and we take up the discussion again when we turn to solids in Chapter 5.

Summary

1   A *homonuclear diatomic molecule* consists of two identical atoms, a *heteronuclear diatomic molecule* consists of two different atoms, and a *polyatomic molecule* consists of any number of atoms greater than two.

2   The *criterion for bond formation* is that an energy decrease should accompany a reorganization of electrons.

3   Bonds may be classed into *ionic* (complete electron transfer), and *covalent* (electron sharing); a *dative bond* is a covalent bond in which one atom provides both electrons.

4   An *ionic bond* may form between two atoms when one atom has a low ionization energy and the other has a high electron affinity.

5   The bonding tendency of atoms is satisfied once they have attained *noble gas, closed-shell configurations* ('the *stable octet*') because their ionization energies are then too high and their electron affinities too low for there to be further energetic advantage in transferring electrons.

6   The strength of a *covalent bond* can be ascribed to the accumulation of the electrons in the internuclear region, where they act as an 'electrostatic glue'.

7   The basic *shape* of a polyatomic molecule is predicted on the basis that the electron pairs minimize their repulsions by moving to the greatest mutual separations.

8   The *valence shell electron pair repulsion* (*VSEPR*) description of molecular structure develops this point by stating that lone pairs of electrons dominate the repulsions between electron pairs in molecules.

9   The *dissociation energy* is the energy required to break a bond.

10 *Single bonds* occur when two atoms share one pair of electrons, *double bonds* when they share two pairs, and *triple bonds* when they share three pairs.

11 A double bond is *weaker* than two single bonds because the electron pairs are in less favourable bonding positions. It is *torsionally rigid*.

12 A triple bond is weaker than three single bonds, and the structure X—C≡C—Y is linear.

13 The stability of *benzene* is due to electron *delocalization*, which reduces electron–electron repulsions and allows the molecule to adopt a regular planar hexagonal shape. The extra stability is called *delocalization stabilization* (or *resonance stabilization*).

14 *Ionic bonding* is non-directional and leads to the formation of extensive aggregates.

15 *Covalent bonding* normally leads to well-defined, individual geometrical structures.

16 A *polar bond* is a covalent bond in which the electrons are shared unequally, resulting in a bond *dipole moment*.

17 A measure of the relative attracting power of an atom for a shared pair of electrons is its *electronegativity*: atoms with high electronegativities (those on the right of the periodic table) attract electrons most strongly and form the negative end of a dipole.

18 A *metal* is an ordered array of positive ions in a mobile sea of electrons.

Problems

1   Write an essay on the modern theory of chemical bonding. Include an account of ionic and covalent bonding, the use and justification of dot-and-cross diagrams, and the role of electron-pair repulsions.

2   List the factors that favour the formation of *ionic* compounds.

3   Account for the tendency of atoms to lose or gain electrons until they have reached a closed-shell configuration.

4   Account for the bonding in (*a*) H_2, (*b*) N_2, (*c*) O_2, (*d*) Cl_2 in terms of both dot-and-cross diagrams and electron distributions.

5   Use dot-and-cross diagrams to account for the bonding in the following compounds: (*a*) KF, (*b*) CaO, (*c*) H_2S, (*d*) PCl_3.

6   On the basis of electron-pair repulsion (VSEPR) theory, predict the shapes of the following molecules: (*a*) H_2S, (*b*) SO_3, (*c*) PH_3, (*d*) PCl_5, (*e*) ClF_3.

7   Explain the nature of the bonding in (*a*) C_2H_4, (*b*) C_6H_6, (*c*) NO_3^-, (*d*) CH_3COOH, (*e*) CH_3COO^-.

8   The polymer PVC, poly(vinyl chloride), is manufactured from the monomer $H_2C=CHCl$, chloroethene ('vinyl chloride'). Describe the bonding in the monomer, and predict its shape.

9   Ethyne ($HC\equiv CH$, 'acetylene') burns with a very hot flame and is used in oxyacetylene welding. Comment on the nature of the bonds in the molecule and account for the large quantity of energy evolved when it is burnt in oxygen.

10   Explain why the electronegativities of atoms affect the extent to which bonds are polar. Which member of the following pairs of molecules has the larger dipole moment: (*a*) HF, HCl; (*b*) CO_2, CS_2; (*c*) NH_3, PH_3?

11   Which of the following molecules have dipole moments: (*a*) CO_2, (*b*) NO_2, (*c*) HBr, (*d*) PCl_3, (*e*) BF_3, (*f*) C_6H_6, (*g*) $C_6H_5CH_3$? Suggest directions for the dipoles.

12   The balance between the abundances of metal ions is essential to the proper functioning of the nervous system, and their transfer through cell membranes depends on their sizes. Account for the fact that the order of sizes is: Mg^{2+} smaller than Ca^{2+} smaller than Na^+ smaller than K^+. (We develop this point in Chapter 8.)

13   (*a*) The atomic radii of sodium and magnesium are 0.157 nm and 0.136 nm, respectively. Why is the magnesium atom smaller than the sodium atom?

   (*b*) The *ionic* radius of sodium is 0.095 nm. The *atomic* radius of neon is 0.160 nm. Comment on the difference between these figures.

(Welsh part question)

14   Describe the shape of each of the following molecules and account for each shape by considering the electronic repulsions within each molecule.

   $SiCl_4$ PCl_3 PF_5 H_2S

(JMB part question)

15   (*a*) What is understood by the term *electronegativity* as applied to an element?

   (*b*) Say how the electronegativities of the elements change
      (*i*) as Group I (alkali metals) is descended;
      (*ii*) as group VII (halogens) is descended;
      (*iii*) as Period II (lithium to fluorine) is crossed.
Comment on your answers to (*i*) and (*ii*) in the light of the electronic configuration of these elements.

   (*c*) Why do the transitional elements, of any one series, have similar electronegativity values?

   (*d*) In what way is the *type* of chemical bond formed between two atoms related to their electronegativities?

Having made a general statement, illustrate your answer by referring to (i) chlorine, (ii) magnesium chloride, (iii) hydrogen chloride. Draw a simple bond diagram in each case.

(e) (i) Using bond diagrams for water and hydrogen chloride, show how hydrogen chloride gas becomes ionized when dissolved in water.

(ii) $ICl + H_2O \rightarrow IOH + HCl$

$ICl + H_2O \rightarrow ClOH + HI$

Which of the two equations above is more likely to represent the hydrolysis of iodine(I) chloride?

Give reasons for your answer.

(SUJB)

16 (a) The bond lengths (in nanometres) and also the dipole moments (in Debye units), of the gaseous hydrogen halides are shown below.

	HF	HCl	HBr	HI
Bond length	0.092	0.127	0.141	0.161
Dipole moment	1.91	1.05	0.80	0.42

(i) What is meant by 'dipole moment'?

(ii) Comment briefly on the reasons for the decrease in dipole moment from HF to HI.

(b) (i) State whether the following molecules would have a dipole moment or not, giving reasons.

Molecule	Dipole moment (Yes or no)	Reason
CH_3Cl		
CCl_4		
BCl_3		
NH_3		

(ii) CO_2 has no dipole moment, while SO_2 has quite a large one. What difference in structure does this suggest?

(c) Many of the differences between water and hydrogen sulphide are ascribed to the fact that the former has hydrogen bonds but the latter has not.

(i) Explain what is meant by "hydrogen bonds".

(ii) List three differences between water and hydrogen sulphide which can be ascribed to hydrogen bonding in the former.

(London)

17 Interatomic bonds are of three ideal types – electrovalent, covalent and metallic.

Describe these types and distinguish between them using the following information:

(i) potassium conducts electricity in the solid state and in the molten state;

(ii) potassium chloride is a crystalline solid of melting point 776 °C; the solid does not conduct electricity, but the molten substance does;

(iii) chlorine has a boiling point of −34 °C; it does not conduct electricity.

(AEB 1977 part question)

18   Discuss how the properties of *four* of the following substances reflect their structure and bonding:

anhydrous aluminium chloride, benzene, caesium chloride, copper(II) sulphate pentahydrate crystals, ice, quartz, poly(propene) (polypropylene), silk.

(Nuffield)

19   Show how knowledge of the spatial distribution of electrons in atoms may be used to explain the bonding in the following molecules:

N_2, O_2, NO, BF_3, SF_4, CO_2.

(Oxford Entrance)

3

Investigating molecules: the determination of structure

In this chapter we see how mass spectrometry is used to identify species, to measure their molar masses, and to determine their molecular formulae and structures. We see how spectroscopy can be used to measure bond lengths and bond angles, and how it can be used to identify molecules. We also see how X-ray diffraction, one of the most useful techniques of modern chemistry, is used to identify molecules and determine their structures.

Introduction

Modern techniques allow us to discover the arrangement of atoms in molecules as complicated as the ones that control biological functions, such as proteins, enzymes, and nucleic acids, and molecular biology could not have developed without the methods we describe here. The future development of modern materials, such as semiconductors, synthetic polymers, and constructional materials like cement and concrete, would be excessively slow, and very hit and miss, if these modern techniques were not available. Spectroscopy and diffraction are also among the most important (and sensitive) techniques available for chemical analysis. They can be used to study complex material in accessible places, such as proteins, and to study simple material in inaccessible places, such as the small molecules present in interstellar space and planetary atmospheres.

3.1 Mass spectrometry

The principles behind the mass spectrometer were described in Chapter 1, where we saw how it was used to measure atomic masses. The spectrometer is sensitive to the mass-to-charge ratio, m/e, of species, and depends on the way that electric and magnetic fields affect the paths of positive ions. A plot of signal strength against magnetic field is a series of spikes corresponding to the values of m/e in the ionized sample. In molecular mass spectrometry the sample is ionized, and this usually results in the molecules falling apart into fragments. The technique therefore involves smashing the molecule into pieces, identifying them, and arriving at the structure of the original molecule by accounting for all the fragments.

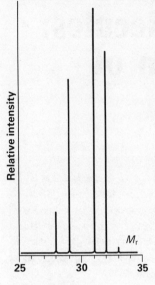

3.1 The mass spectrum of methanol

There are simplifying features in the technique as well as complications. In the first place the ionizing electron beam is usually not energetic enough to remove more than one electron from each sample molecule, and so we rarely need to worry about the presence of multiply-charged ions of the form A^{2+}, A^{3+}, *etc*. The mass spectrum of A therefore consists of peaks corresponding to the m/e ratio for the *primary ion* A^+ together with whatever singly charged species it falls apart into (but sometimes the primary ion A^+ is so unstable that only its fragments are detected).

As an example of the type of information obtained consider the case of methanol, CH_3OH. Its mass spectrum is shown in Fig. 3.1. Since $M_r = 32$ we might expect a single peak at an m/e ratio corresponding to $M_r = 32$ (colloquially this is referred to as 'a peak at $m/e = 32$') due to the singly ionized molecule $(CH_3OH)^+$. The primary ion is certainly in the spectrum, but there are also others at m/e ratios corresponding to $M_r = 31$, 29, and 28. These are consistent with the presence of the fragments CH_3O^+, CHO^+, and CO^+, respectively.

Molecules of similar constitution may give rise to quite different mass spectra. This is illustrated by the spectra of two isomeric materials of molecular formula $C_4H_{10}O$: the mass spectrum of butan-1-ol is shown in Fig. 3.2(*a*) and is seen to be quite different from the spectrum of 2-methylpropan-2-ol shown in Fig. 3.2(*b*). The explanation is that the two species break up in different ways when smashed by an energetic electron beam. This is turned to advantage in modern mass spectrometry because spectra can be interpreted by referring to a library of patterns kept in a computer. Each species can be identified by taking its *fragmentation fingerprint*. As in all branches of science, it is sensible to use a variety of techniques to tackle a problem, and mass spectrometry is normally employed in conjunction with the other methods we describe later.

Mass spectrometry is used to determine the *molecular formula* of a sample. At first sight there is a problem which might appear to make it

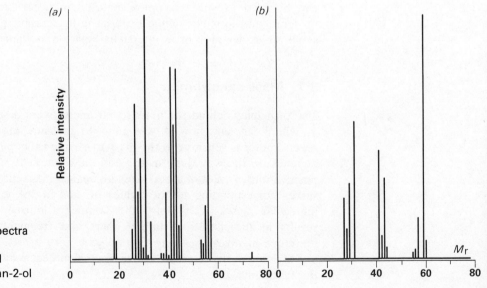

3.2 The mass spectra of the isomers (*a*) butan-1-ol and (*b*) 2-methylpropan-2-ol

Principles of physical chemistry

completely useless except in the case of very small molecules. Take for instance a molecule found to have $M_r = 194$. Its molecular formula could be $C_{10}H_{18}N_4$, or $C_8H_6N_2O_4$, or $C_{10}H_{10}O_4$, or any one of about another 35 combinations of C, H, N, and O (not to mention dozens of other possibilities if the presence of other elements is suspected). This problem is solved by building spectrometers with such high resolution that they can distinguish very small differences of mass. Modern spectrometers can resolve m/e ratios to better than one part in a hundred thousand. The reason why this overcomes the identification problem is that the molecules mentioned above do have different M_r values when they are calculated more precisely. Thus $M_r(C_{10}H_{18}N_4) = 194.1531$, $M_r(C_8H_6N_2O_4) = 194.0328$, and $M_r(C_{10}H_{10}O_4) = 194.0579$, and so a resolution of 1 part in 10 000 is sufficient to distinguish them. The identification can then be confirmed by seeing whether a peak is accompanied by another arising from the presence of molecules containing a naturally occurring isotope. Thus, if there is a $C_8H_6N_2O_4$ peak with the normal ^{12}C isotope, there also ought to be a peak with one of the ^{12}C atoms replaced by ^{13}C (which occurs naturally in 1.1% abundance). Figure 3.1 has a low intensity peak at m/e corresponding to $M_r = 33$; this is due to $(^{13}CH_3OH)^+$ from the $^{13}CH_3OH$ present in 1.1% abundance in ordinary methanol.

Apart from its use in analysis, mass spectrometry has many other applications. For instance, it is used for quality control in steel making. Also, small mass spectrometers tuned to m/e corresponding to $M_r = 4$ are extremely sensitive leak detectors in vacuum systems. The outside of the apparatus is sprayed with helium (mass number 4), and if any atoms are sucked in they give rise to a signal in a small spectrometer attached to the system. This method can detect leaks of down to 10^{-12} cm^3 s^{-1} of gas at 1 atm pressure, a leakiness that in many systems is negligible. (At that rate of leakage, a bicycle tyre would stay inflated for about ten million years!) Mass spectrometers are also used to follow the course of chemical reactions, because some of the reaction mixture can be drawn off continuously and analyzed both for the components present and, from the intensities of the peaks, the time variation of their concentrations. Mass spectrometers are frequent passengers on planetary missions. The Viking spacecraft carried a mass spectrometer to Mars, and sent back a detailed analysis of the composition of its atmosphere (which is 95% carbon dioxide, 2.5% nitrogen, 1.5% argon, and 1% other gases).

3.2 Molecular spectroscopy

The principles of molecular spectroscopy are the same as those of atomic spectroscopy. The basic equation is the *Bohr frequency condition* for the frequency ν of the light emitted (or absorbed) when a molecule changes its energy from E(upper) to E(lower) or *vice versa*:

$$h\nu = E(\text{upper}) - E(\text{lower}). \tag{3.2.1}$$

The basis of this expression was described in Section 1.3. When a molecule drops to a lower energy its excess energy is carried away as a photon of light. In an absorption step the photon must bring the necessary energy if the molecule is to climb to a higher energy state (that is, make a *transition* to a higher energy state). The difference between the spectra of atoms and molecules is that while changes of the energies of atoms arise only from the changes in the distributions of their electrons, in molecules energy is also required to make a molecule rotate more rapidly or to vibrate more vigorously, and is discarded when rotations and vibrations decelerate. In other words, molecular energies can be changed in three ways: by shifts of their electrons, by changes in their vibrations, and by changes in their speeds of rotation.

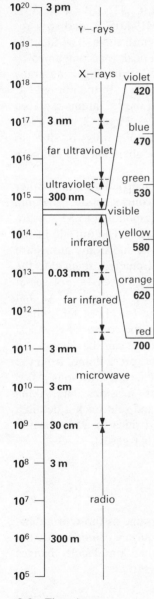

Example: Calculate the change in energy that must occur when a molecule emits a photon of blue (450 nm) light.

Method: Use eqn (3.2.1) with $h = 6.626 \times 10^{-34}$ J s and relate wavelength to frequency through $\nu = c/\lambda$, where c is the speed of light, $c = 2.998 \times 10^8$ m s^{-1}.

Answer: The frequency of the light is

$$\nu = \frac{2.998 \times 10^8 \text{ m s}^{-1}}{450 \times 10^{-9} \text{ m}} = 6.662 \times 10^{14} \text{ s}^{-1} \ (6.662 \times 10^{14} \text{ Hz}).$$

The energy change required is

$$E(\text{upper}) - E(\text{lower}) = (6.626 \times 10^{-34} \text{ J s}) \times (6.662 \times 10^{14} \text{ s}^{-1})$$
$$= 4.414 \times 10^{-19} \text{ J}.$$

Comment: 4.414×10^{-19} J may not seem like very much energy, but when expressed in terms of the energy per mole it corresponds to $(6.022 \times 10^{23} \text{ mol}^{-1}) \times (4.414 \times 10^{-19} \text{ J}) = 265.8$ kJ mol^{-1}, a very significant quantity.

The energy changes due to molecular rotation and vibration are very much smaller than those due to shifts of electrons. As a result, the absorption frequencies arising from vibrational and rotational changes lie at much lower frequencies (longer wavelengths). The electromagnetic spectrum, together with the names given to its various regions, is set out in Fig. 3.3. We shall see that whereas changes of electron distribution (*electronic transitions*) are responsible for absorptions in the high energy *visible and ultraviolet regions* of the spectrum, vibrational changes are responsible for absorption in the *infrared region* and rotational changes for absorption in the *microwave region*.

Molecular spectra are normally observed in absorption, and the layout of a typical spectrometer for spectra in the visible, ultraviolet, or infrared regions (microwave techniques tend to be more specialized) is illustrated in Fig. 3.4. Radiation from an appropriate source is passed through a prism or a diffraction grating, and the selected *monochromatic* (single

3.3 The electromagnetic spectrum

Principles of physical chemistry

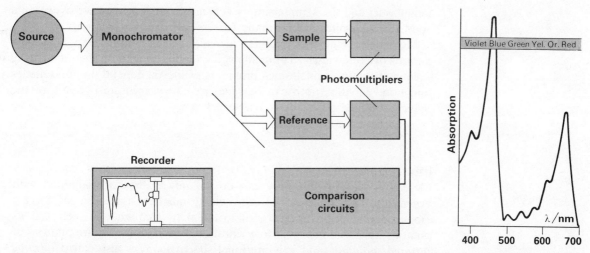

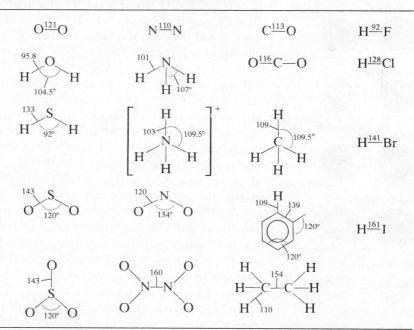

3.4 (above) The layout of a typical double-beam spectrometer

3.5 (above right) The absorption spectrum of chlorophyll in the visible region

frequency) component is then divided into two; one beam is passed through a sample and the other is passed through a blank. The difference in the intensities is measured, and plotted as the absorption spectrum. A typical spectrum (of chlorophyll) is shown in Fig. 3.5.

Microwave spectra

When microwaves pass through samples they stimulate molecular rotation. A gas has to be used for quantitative work because then molecules can rotate freely. In a liquid they jump round and are jostled by their

Table 3.1 Some molecular dimensions

Note: The nature of bonds (single, double, *etc.*) has not been specified. Bond lengths are in picometres ($1 \text{ pm} = 10^{-12}$ m)

neighbours and the sample simply gets hot (as in a microwave oven). In a solid the molecules are normally frozen into fixed orientations.

Molecular rotational energies are *quantized*; that is, can take only various discrete values. These allowed energies depend on the *moments of inertia* of the molecule. Since moments of inertia depend on bond angles and bond lengths, both can be measured very accurately. Some of the values obtained are listed in Table 3.1.

Infrared spectra

Molecules vibrate, but they can do so only if they are supplied with exactly the right energy. Therefore electromagnetic radiation can force a molecule into more energetic vibration if its frequency is such that its photons have that energy. Most vibrational excitations require photons of *infrared radiation*, and so vibrational spectroscopy is also called *infrared spectroscopy*. When you get warm in front of a radiant heater (or the sun) you are behaving like a vibrational spectroscopy sample, and the molecules in your body capture the infrared radiation and vibrate more energetically.

Atoms swing around like particles attached to springs. Just as the natural frequency of a particle on a spring depends on its mass and the stiffness of the spring, so the natural frequencies of atoms depend on their masses and the stiffness of the bonds connecting them to the rest of the molecule. Hydrogen atoms are very light, and so usually vibrate at high frequencies. For instance, the C—H stretching and squashing motion (this is called a 'C—H stretch') occurs at a frequency of 8.7×10^{13} Hz while C—Cl vibrates at about 2.1×10^{13} Hz. A double bond is stiffer than a single bond between the same two atoms, and so it can be expected to vibrate at a higher frequency. This is generally found in practice, and while C—C vibrates at about 3.0×10^{13} Hz, C=C vibrates at 5.0×10^{13} Hz, and C≡C vibrates at about 6.6×10^{13} Hz. A bond is generally less rigid to bending than to stretching, and so bending vibrations generally occur at lower frequencies than stretching vibrations. For example, whereas the C—H stretch vibrates at around 8.7×10^{13} Hz, the C—H bending motion (the 'C—H bend', when the hydrogen atom wags from side to side) vibrates at about 4.2×10^{13} Hz. Groups of atoms like —CH₃ vibrate at characteristic frequencies and their presence can be detected by noting the corresponding absorptions in the infrared spectrum of a sample. The vibrational spectrum is therefore a kind of fingerprint of the groups present. For example, the benzene ring has a symmetrical swelling and contracting motion (a 'breathing mode') at 3.0×10^{13} Hz, and detecting an absorption at that frequency is a very good indication of the presence of the aromatic ring. Figure 3.6 shows some of the vibrational *modes* of a group of atoms in a molecule. Each one has a characteristic frequency. This *fingerprinting* capability of infrared spectroscopy is immensely useful for the identification of complex organic molecules. Some typical vibrational modes and the wavelengths of the radiation needed to excite them, are listed in Fig. 3.6.

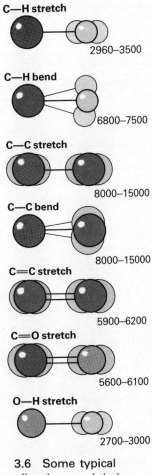

C—H stretch
2960–3500

C—H bend
6800–7500

C—C stretch
8000–15000

C—C bend
8000–15000

C=C stretch
5900–6200

C=O stretch
5600–6100

O—H stretch
2700–3000

3.6 Some typical vibrations and their wavelengths (λ/nm)

Principles of physical chemistry

Example: An infrared spectrum is shown in Fig. 3.7. Deduce the identity of the sample.

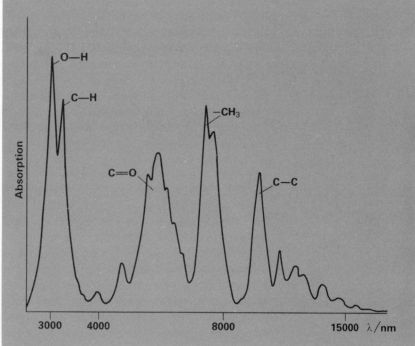

3.7 The infrared absorption spectrum of ethanoic acid

Method: Refer to Fig. 3.6, which lists typical infrared absorption wavelengths, and identify some of the groups present.

Answer: Five peaks can be identified readily. The peaks at 3000 nm correspond to O—H and C—H stretches. The group at 6000 nm indicates a C=O group. The group at 8000 nm is due to a —CH₃ group, and the 10 000 nm vibration is the slow vibration typical of C—C. Therefore we deduce the presence of CH₃, C—C, C=O, O—H, and C—H; hence the sample is probably ethanoic acid.

Comment: The C=O stretch dominates the infrared spectra of aldehydes and ketones, and makes them easy to identify in a sample. There is much more information in the spectrum than we have extracted, but its complete analysis is a very complex and skilled task.

Ultraviolet and visible spectra

Photons of visible light carry a lot of energy (about $200 \, \text{kJ} \, \text{mol}^{-1}$ for yellow light – a 100 W sodium lamp generates 1 mol of photons in half an hour – and $260 \, \text{kJ} \, \text{mol}^{-1}$ for blue), and ultraviolet photons carry even more. When they strike a molecule they can cause shifts of electrons from one region to another. In the process they are absorbed. For instance, the spectrum of chlorophyll in Fig. 3.5 shows that the molecule can absorb in the red and in the blue; as a consequence, the photons corresponding to

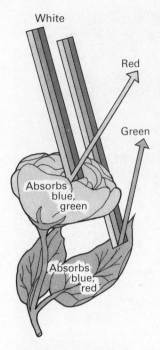

White

Red

Green

Absorbs blue, green

Absorbs blue, red

3.8 Objects appear coloured as a result of selective absorption

these two colours are extracted from sunlight, and the reflected light is green, Fig. 3.8.

Absorption in the visible and ultraviolet regions is often a result of the presence of particular groups of atoms in the molecule. These groups are called *chromophores* (from the Greek for 'colour bringer'). For instance, the carbonyl group, $\diagup C{=}O$, has an absorption maximum at $\lambda = 185$ nm, and the carbon-carbon double bond $\diagup C{=}C\diagdown$ has one at $\lambda = 170$ nm. Complex ions containing transition metals often have characteristic colours; the $[Cu(H_2O)_6]^{2+}$ ion in aqueous copper(II) sulphate is blue on account of its absorption in the red region of the spectrum. It follows that ultraviolet and visible absorption spectra can be used to identify the presence of groups in a compound; in some cases the identity of the compound as a whole can be decided.

The electronic spectra of molecules have many applications. The analytical use, the identification of groups and molecules, has already been mentioned. Measuring the intensity of absorption is also a very useful way of determining concentration.

Information from electronic absorption spectra is used to understand the chemical reactions brought about by light. The initial step in these *photochemical reactions* is the absorption of a photon of light by a molecule, which is followed by the reaction of the molecule as a result of its high energy. Photochemical reactions are essential to all aspects of life (even to things like men, women, and fungi, which do not photosynthesize). Not only do they capture the radiant energy of the sun, the first step in the sequence of events that makes life possible, but they are of increasing importance as an energy source. Photochemical reactions also play many roles in everyday affairs, such as in the fading of dyes, sunbathing, and photography. The absorption in the ultraviolet region by p-aminobenzoic acid ($NH_2 \cdot C_6H_4 \cdot COOH$) is the basis of some sun-tan creams; 'whiter-than-whiteness' with detergents is obtained by increasing the optical brightness of clothes by using additives that absorb in the ultraviolet and re-emit in the visible.

3.3 X-ray diffraction

X-ray diffraction is the technique used to obtain the kind of diagram shown in Fig. 3.9. It gives a vivid picture of the structure of the molecule (anthracene in this case). The lines on the diagram are contours of equal electron density, and so not only can bond angles and lengths be measured but detailed information about the variation of the electron density throughout the molecule can be obtained. The technique has found its most exciting applications in molecular biology. X-ray diffraction is the principal source of our understanding of the structure of proteins, enzymes, and nucleic acids. *X-ray crystallography* is the application of the technique to the study of crystal structure; we meet it in Chapter 5.

The basis of the diffraction technique is the interference between waves. If the crests of two or more waves coincide, then there is an increase of amplitude, but if a crest coincides with a trough, Fig. 3.10,

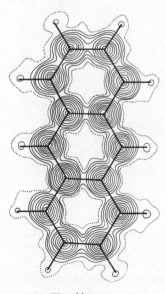

3.9 The X-ray diffraction picture of anthracene

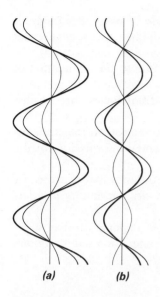

3.10 (a) Constructive and (b) destructive interference of waves

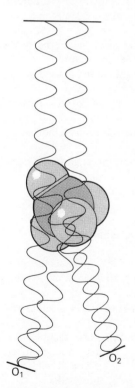

3.11 Scattering from different parts of a molecule, and the interference at two points

there is a cancellation. In the present case the waves used lie in the X-ray region of the electromagnetic spectrum, with wavelengths of about 100 pm (0.1 nm), because only wavelengths this short can show up details of molecular structure.

The interference can be used to explore the structure of a molecule. Figure 3.11 shows what happens to a beam of X-rays when it is directed on to a molecule. The radiation is scattered from the molecule, and an observer sees it coming from many directions. The illustration shows two of the scattered waves. When the detector is positioned at O_1, Fig. 3.11, the crests are coincident, and so it registers a strong X-ray intensity. When it moves to O_2 the scattered waves travel different distances, and the crests of one coincide with the troughs of the other. As a result the detector records zero X-ray intensity. By measuring the intensity of radiation at all orientations around the molecule, taking into account the scattering from every region and not just the two we have shown, the observer records a very complicated intensity distribution. When it is analyzed (always on a computer because the calculations are so complicated) the output is the kind of diagram shown in Fig. 3.9.

Electrons accelerated to high speeds behave like waves with wavelengths of about 100 pm (the wave-particle duality was mentioned in Chapter 1), and so can be used in diffraction experiments. *Electron diffraction* is used to study molecules in the gas phase, because the beam does not penetrate solids. Neutrons from reactors also have suitable wavelengths and are used in *neutron diffraction* studies of solids. The most important technique, however, remains X-ray diffraction, and nothing else has done more to make the invisible open to inspection and measurement.

Summary

1 *Mass spectrometry* is used to determine the masses of molecules, their molecular formulas (at high resolution), and their constitution (from their fragmentation patterns).
2 The *mass spectrometric technique* is sensitive to the mass-to-charge ratio, m/e, of positive ions produced either directly by electron bombardment or by subsequent fragmentation.
3 Molecular spectra can be explained in terms of the *Bohr frequency condition*, eqn (3.2.1).
4 *Electronic, rotational*, and *vibrational energy levels* of molecules are *quantized*.
5 Absorption in the *microwave region* of the electromagnetic spectrum can be accounted for in terms of the stimulation of *rotational excitation* of gas-phase molecules.
6 *Microwave spectra* can be interpreted in terms of the *moments of inertia* of molecules, and hence in terms of their geometry, their *bond lengths* and *bond angles*.
7 *Infrared absorption spectra* are due to the stimulation of molecular *vibrations*.

8   Infrared spectra are used for the *identification* of molecules in gases, liquids, and solids (mainly in liquids), and for finding the *strengths and stiffnesses* of their bonds.

9   *Ultraviolet and visible absorption spectra* are due to the stimulation of *electronic transitions* by the incident radiation.

10 Groups that absorb at characteristic frequencies are called *chromophores*.

11 *Electronic spectroscopy* is used to obtain information on the strengths of bonds, to identify species present, and as essential information for the study of photochemical reactions.

12 *X-ray diffraction* is based on *interference* between waves scattered by different parts of a molecule. It gives details of *electron density distributions*.

13 *Electron diffraction* and *neutron diffraction* depend on the wave characteristics of fast electrons and neutrons, and are used to obtain structural information.

Problems

1   Write an essay outlining the principles of spectroscopy. Cover the basic process, account for absorption and emission spectroscopy, explain how different regions of the spectrum arise from excitation of different transitions, and mention some applications.

2   Explain the usefulness of the fragmentation pattern in mass spectrometry.

3   Suggest the form of the mass spectrum of (*a*) methane, (*b*) water.

4   What extra peaks would be expected on the basis that the methane used in the last question contained 1% ^{13}C?

5   How might deuteration of methanol (the substitution of one proton by one *deuteron*, $^2_1H^+$) affect its mass spectrum?

6   Predict the form of the mass spectrum from gaseous molecular chlorine of natural isotopic abundance.

7   Which spectroscopic technique would be suitable for the investigation of each of the following problems?

   (*a*) The presence of trace amounts of arsenic in the blood stream.

   (*b*) The composition of the atmosphere of Venus.

   (*c*) The nature of exhaust gases from a car engine.

   (*d*) The vibrational frequency of the O—H bond in ethanol.

   (*e*) The bond length of HCl.

   (*f*) The purity of ethanol.

8   Explain the terms *chromophore, vibrational mode, stretching vibration, breathing mode*.

9   Suggest what will happen to the frequency of vibration of C—H on deuteration.

10 Describe any ONE spectrometric method for the determination of molecular structure (possible methods include infrared spectroscopy or mass spectrometry but NOT diffraction methods). For the method of your choice you should deal with the practical procedure AND with the interpretation of the results.

(Nuffield)

4

Free particles: the nature of gases

In this chapter we look at gases, the simplest form of matter. We see how experiments led to the idea of the 'perfect gas' and to the perfect gas equation and how the basic ideas are readily extended to mixtures. The picture of a gas as a collection of chaotically moving particles provides an explanation of the perfect gas law. It can also be used to relate the speeds of the particles to the temperature. The perfect gas law is only approximately applicable to real gases under normal conditions, and so we also examine why real gases behave differently.

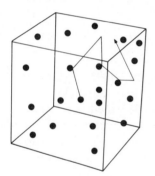

4.1 The nature of a gas (all particles move)

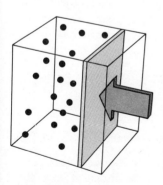

4.2 Gases are easy to compress

Introduction

The name *gas* comes from the word chaos. That summarizes very well the principal feature of this the simplest form of matter. We picture a gas as a swarm of particles moving randomly and chaotically, constantly colliding with each other and with the walls of the containing vessel, Fig. 4.1. The pressure exerted by the gas is a result of the never-ending battering of the particles on the walls. The tendency of any gas to fill the available volume is a result of the freedom that the particles have to explore everywhere open to them. The easy compressibility of gases is due to the presence of so much empty space between their particles, Fig. 4.2.

The chaotic nature of gases does not mean that they cannot be discussed scientifically. In fact some of the earliest experiments in the field that has since become physical chemistry were done on gases. They led to the formulation of Boyle's law (in 1662) and Charles's law (in 1787), Box 4.1.

Box 4.1: The development of the understanding of gases

Robert Boyle was born in Ireland in 1627 and went to Oxford in 1654. There he constructed a pump with the help of Robert Hooke, and set about his studies of the properties of gases. His avowed motivation for science was religious, in the sense that he wanted to understand the working of the creation. In the course of his work he attacked Aristotle's views on the constitution of the world in terms of earth, air, fire, and water, and argued in favour of a corpuscular nature of matter. He proposed his law in 1662; it was proposed independently in France by Edmé Mariotte (in 1676) where it bears his name in place of Boyle's. France was the source of further understanding of gases, for Jacques Alexandre César Charles was a Frenchman with a consuming interest in ballooning (the early hydrogen balloons were often referred to as *charlières*), which stimulated him to study the effect of temperature on gases. He did not publish his findings and they were re-discovered by Gay-Lussac, another Frenchman, who did. The study of real gases and their liquefaction was put on secure foundations by the Irishman Thomas Andrews, and the first largely successful attempt to account for their properties was made by the Dutchman J. D. van der Waals in 1873. His was the first attempt to establish an *equation of state* for a real gas, but his successors have suggested another 100 or so ways of describing them in different ways. Modern work concentrates on two aspects of gases. One is the description of their imperfections in terms of the *virial equation* of state (which is normally associated with another Dutch physical chemist, H. Kamerlingh Onnes, who proposed it in 1901) and the other concerns the dynamical properties of flowing gases, such as their viscosities and the way they conduct heat.

4.1 Looking at gases: the gas laws

There are three characteristic properties of gases: their response to pressure, their response to temperature, and their dependence on amounts. Many gases also have characteristic smells: smell is related to the ability of molecules to attach to receptors in the membranes of the nose. They depend on the structures of the molecules (*e.g.*, their shape, the presence of lone pairs, and hydrogen bonding). In this chapter we concentrate on the physical properties.

The pressure dependence

Robert Boyle measured the volume of a sample of gas subjected to different pressures. He noticed that the higher the pressure the smaller the volume, and found that the volume is *inversely proportional* to the applied pressure, Fig. 4.3:

> *Boyle's law: At constant temperature and for a fixed mass of gas,*
> $V \propto 1/p.$ (4.1.1)

An equivalent statement is

> *Boyle's law: pV = constant* (temperature and mass fixed). (4.1.2)

We now know that this law is only an approximation. Nevertheless there are two points of great significance which stop us dismissing it as a 300 year old historical irrelevance.

In the first place Boyle's law is very useful even though it is an

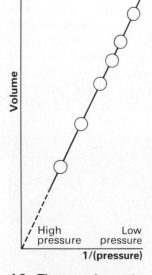

4.3 The experimental evidence for Boyle's law

Principles of physical chemistry

approximation. It gives a simple way of predicting the volume of a gas when it is subjected to pressure, and the deviations from predictions are important only when the gas is dense or cold. The second point is much more significant. Careful measurements have shown that as the density of a sample of any gas is reduced, so Boyle's law is obeyed increasingly accurately. The law is in fact a statement about an ideal fluid called a *perfect gas* or an *ideal gas*, and is obeyed exactly when the gas is infinitely diffuse.

Example: A four-cylinder $1300\,cm^3$ engine has a compression ratio of $9.0:1$. What is the volume remaining in one cylinder when the piston is at the top of its stroke?

Method: This is an application of Boyle's law, eqn (4.1.2), in the form $(pV)_{\text{top of stroke}} = (pV)_{\text{bottom of stroke}}$. The *compression ratio* is the ratio of the pressures at the top and bottom of the stroke, the temperature being constant. The volume at the bottom of the stroke is one quarter of the total specified volume of the four-cylinder engine.

Answer: Volume of one cylinder $V_{\text{bottom}} = \frac{1}{4} \times (1300\,cm^3) = 325\,cm^3$. Ratio of pressures $p_{\text{top}}/p_{\text{bottom}} = 9.0$. Application of Boyle's Law then gives

$$V_{\text{top}} = \frac{(pV)_{\text{bottom}}}{p_{\text{top}}} = \frac{325\,cm^3}{9.0} = 36\,cm^3.$$

Comment: In a working engine the compression stroke also heats the gas. We shall see how to take this into account next. It is important to realize that Boyle's law applies specifically to a compression or an expansion at constant temperature.

The temperature dependence

The experiments of Gay-Lussac and Charles led them to the conclusion that a gas responds to temperature in a very simple way:

Charles's law: the volume of a gas is proportional to the temperature, the pressure and mass being fixed.

More recent observations have shown that *real gases* (that is, actual gases) deviate from this law too. Nevertheless, under normal conditions the deviations are small, and disappear in the extreme case of an infinitely diffuse gas.

Since, according to Charles's law, the volume of a gas is proportional to the temperature, it follows that at a sufficiently low temperature the volume should become zero. Experimentally it is found that the volumes of any amount of any gas when plotted against temperature, Fig. 4.4, extrapolate to zero at the same temperature. That temperature is $-273.15\,°C$. (Once again this is strictly true only when the gases are so diffuse that they are behaving perfectly.) This suggests that there is a natural *absolute zero of temperature*, and the natural temperature scale, or

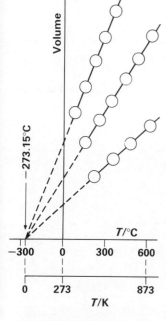

4.4 The experimental evidence for Charles's law (different samples correspond to different masses of gas)

Kelvin scale, takes $T = 0$ as the temperature at which a perfect gas occupies zero volume.

Once the Kelvin scale of temperature is available, Charles's law can be expressed very simply as

$$V \propto T, \text{ pressure and mass constant, } T \text{ on the Kelvin scale.} \qquad (4.1.3)$$

The dependence on amount

The third piece of information is due to Avogadro. A 2 g sample of hydrogen gas is found to occupy the same volume as a 32 g sample of oxygen (which at 0 °C and atmospheric pressure is about 22.4 dm^3). We know that 2 g of hydrogen corresponds to 1 mol of H_2 and that 32 g of oxygen corresponds to 1 mol of O_2; hence we are led to

Avogadro's law: equal amounts of substance of gases occupy the same volume at the same temperature and pressure.

Therefore the volume occupied by unit amount of substance,[1] the *molar volume*, is the same for all gases at the same temperature and pressure. If the amount of substance of gas is n, and its molar volume is V_m, then the volume of the sample is

$$V = nV_m. \qquad (4.1.4)$$

The number of particles per unit amount of substance (*e.g.*, the number per mole) is called *Avogadro's constant, L*. It has the value

$$L = 6.022 \times 10^{23} \text{ mol}^{-1}.$$

Note that L is a *constant*, not a number (it has dimensions). There are several ways of determining L, and we meet one in the next chapter.

---

[1] *Amount of substance* (*e.g.*, so many moles of something) is a quantity described in more detail in *Background information* at the end of the book.

Example: What volume is occupied by 2.70 g of hydrogen gas, given that the molar volume of a perfect gas is 22.4 dm^3 mol^{-1} under the same conditions of pressure and temperature?

Method: Assume that hydrogen behaves perfectly and use Avogadro's law in the form $V = nV_m$, where n is the amount of substance of the gas and V_m its molar volume. Find n on the basis that the molar mass of H_2 is 2.016 g mol^{-1} (*i.e.*, 2×1.008 g mol^{-1}, see inside back cover).

Answer: The amount of substance of H_2 in a sample of mass 2.70 g is

$$n = \frac{2.70 \text{ g}}{2.016 \text{ g mol}^{-1}} = 1.34 \text{ mol}.$$

Since $V_m = 22.4$ dm^3 mol^{-1}, it follows that

$$V = (1.34 \text{ mol}) \times (22.4 \text{ dm}^3 \text{ mol}^{-1}) = 30.0 \text{ dm}^3.$$

Principles of physical chemistry

Comment: According to Avogadro's law, we can also state that 2.70 g of hydrogen gas contains $(1.34 \, mol) \times (6.022 \times 10^{23} \, mol^{-1}) = 8.07 \times 10^{23}$ molecules of H_2. Since the sample occupies 30 dm³, an individual molecule has about $(30 \times 10^{-3} \, m^3)/(8.07 \times 10^{23}) = 3.72 \times 10^{-26} \, m^3$ of space to itself. This corresponds to a cube of side 3.3×10^{-9} m, or 3.3 nm. Since the diameter of the molecule is about 0.2 nm we can begin to visualize the diffuseness of the gas: its nearest neighbour is about 15 molecular diameters away, and so a model of a molecule 2 cm long has its nearest neighbours about 30 cm distant in a gas under the conditions of this *Example*, which are similar to normal atmospheric conditions.

Avogadro's law is an idealization, and is known not to be strictly true. Nevertheless it is a very helpful broad summary of the behaviour of gases under normal conditions, and it is exactly true when the density of the gas is vanishingly small. In other words, it is another property of a perfect gas.

The perfect gas equation

The experimental observations we have described have identified three types of behaviour of a perfect gas. Under the appropriate conditions they are pV = constant, $V \propto T$, and $V \propto n$. All three can be combined into the single, simple expression $pV \propto nT$, or, on writing the coefficient of proportionality as R, as

$$pV = nRT. \tag{4.1.5}$$

This is the *perfect gas equation* (or the *perfect gas law*). The coefficient R, the same for every gas, is the *molar gas constant* (which is usually just called 'the gas constant'). It can be determined once and for all by evaluating $R = pV/nT$ for some gas. When 0.20 g of hydrogen (corresponding to 0.10 mol of H_2) is confined in a 1.0 dm³ vessel of 0 °C (273 K) it is found to exert a pressure of 227 kPa ($2.27 \times 10^5 \, N \, m^{-2}$). Therefore

$$R \approx \frac{(2.27 \times 10^5 \, N \, m^{-2}) \times (1.0 \times 10^{-3} \, m^3)}{(0.10 \, mol) \times (273 \, K)}$$

$$= 8.3 \, N \, m \, K^{-1} \, mol^{-1} = 8.3 \, J \, K^{-1} \, mol^{-1}.$$

As progressively smaller amounts of gas are confined in the vessel at the same temperature, the value of pV/nT approaches a constant. When the value of pV/nT is extrapolated to $n = 0$, it is found that

$$R = 8.314 \, J \, K^{-1} \, mol^{-1}.$$

The same value of R is found whatever the nature of the gas.

The perfect gas equation makes it easy to predict the state of a gas under different conditions. For instance, in some cases it is convenient to express the volume of a gas measured under one set of conditions of temperature and pressure to a standard set of conditions. *Standard temperature and pressure* (s.t.p.) is defined as 0 °C and 1 atm (that is, $T = 273.15$ K and $p = 101.325$ kPa, respectively). Suppose, for example, that the molar volume of a gas is found to be V_m at some temperature T, and pressure p; then the conversion to s.t.p. is

$$V_m(\text{s.t.p.}) = (273.15 \text{ K}/T) \times (p/101.325 \text{ kPa}) \times V_m, \qquad (4.1.6)$$

where we have used eqn (4.1.5) on the assumption that the gas behaves perfectly. The perfect gas law is also the basis of one method of measuring the molar mass of a gas (or vapour). The method involves rearranging the gas law into $1/n = RT/pV$ and making use of the result that if the mass of a sample is M and its molar mass is M_m, then the amount of substance present is $n = M/M_m$ so that $1/n = M_m/M$. Therefore if the volume of a known mass of gas is measured at a known pressure and temperature, the molar mass can be deduced by combining these two expressions for $1/n$ into $M_m = MRT/pV$. In the modern version of the *Victor Meyer method* a known mass of sample is placed in a *gas syringe*, Fig. 4.5. The temperature and pressure are measured, and the volume occupied by the gas is noted. This technique is now used mainly for analysing the composition of mixtures of gases because there are better methods (*e.g.*, mass spectrometry) for measuring the molar masses themselves.

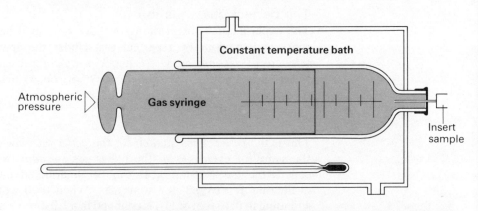

4.5 A gas syringe

Example: An ammonia plant produces 1000 tonne per day by the Haber-Bosch process. The plant operates at 200 atm and 525 °C. What volume of ammonia at that pressure is produced each day? What volume does that correspond to at 1 atm pressure and 25 °C?

Method: Assume that the gas behaves perfectly and apply eqn (4.1.5) in the form $V = nRT/p$ with $R = 8.206 \times 10^{-2} \text{ dm}^3 \text{ atm K}^{-1} \text{ mol}^{-1}$. Convert the temperature to the Kelvin scale and calculate n on the basis that the molar mass of ammonia is 17.0 g mol^{-1} (see inside back cover). For the second part use the gas law in the form $p_1 V_1/T_1 = p_2 V_2/T_2$ for a given amount of gas.

Answer: The temperature of the plant is $T = (273 + 525) \text{ K} = 798 \text{ K}$. The amount of substance of NH_3 corresponding to 1000 tonne ($10^3 \times 10^3 \text{ kg} = 10^6 \text{ kg}$) is

$$n = \frac{10^6 \text{ kg}}{17.0 \text{ g mol}^{-1}} = \frac{10^9 \text{ g}}{17.0 \text{ g mol}^{-1}} = 5.88 \times 10^7 \text{ mol.}$$

The perfect gas law, eqn (4.1.5), then gives

$$V = nRT/p = \frac{(5.88 \times 10^7\,\text{mol}) \times (8.206 \times 10^{-2}\,\text{dm}^3\,\text{atm}\,\text{K}^{-1}\,\text{mol}^{-1}) \times (798\,\text{K})}{(200\,\text{atm})}$$

$$= 1.93 \times 10^7\,\text{dm}^3.$$

The volume corresponding to the same amount of substance but at 1 atm pressure and 25 °C (298 K) is

$$V_2 = (T_2/T_1) \times (p_1/p_2) \times V_1$$

$$= \left(\frac{298\,\text{K}}{798\,\text{K}}\right) \times \left(\frac{200\,\text{atm}}{1\,\text{atm}}\right) \times (1.93 \times 10^7\,\text{dm}^3) = 1.44 \times 10^9\,\text{dm}^3.$$

Comment: These huge volumes give some idea of the large scale of industrial processes: $10^9\,\text{dm}^3$ corresponds to a cube of side 100 m. In real life the ammonia has to be treated as a real gas, and at the pressures dealt with in this *Example* the deviations from perfect behaviour are significant. Nevertheless, the numbers we have arrived at give an idea of the magnitudes involved.

Mixtures of gases

In chemistry we often have to deal with mixtures of gases (the atmosphere itself is an example). The law introduced by Dalton as a result of his experimental studies (recall his interest in meteorology mentioned in Box 1.1) extends the perfect gas law to include mixtures.

The problem can be expressed as follows. Suppose an amount of substance of gas n_A (*e.g.*, 0.1 mol H_2) is introduced into a container of volume V. Then the perfect gas law predicts that the pressure it exerts is $p_A = n_A RT/V$. If another gas B had been introduced instead it would have exerted a pressure $p_B = n_B RT/V$, where n_B is its amount of substance (*e.g.*, 0.2 mol O_2). But suppose the second gas had been introduced into the container that already had the first gas inside. What pressure would each gas exert, and what would be the total pressure?

Dalton's observations led him to conclude that neither gas was affected by the presence of the other. Therefore, even though gas A is present, gas B exerts the pressure p_B given by the expression above. This observation is expressed as follows:

> *Dalton's law: Each gas in a mixture exerts the same pressure as when it alone occupies the container at the same temperature.*

The pressures p_A and p_B exerted by the individual gases are called their *partial pressures*. The *total pressure*, the pressure we observe if there is a pressure gauge attached to the vessel, is the sum of these partial pressures:

$$p = p_A + p_B. \tag{4.1.7}$$

The gas laws provide a simple way of calculating the partial pressure of each component, given the composition of the mixture and the total

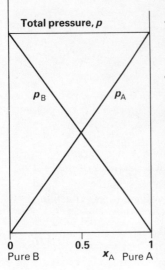

Pressure

Total pressure, p

p_B p_A

0   0.5   1
Pure B x_A Pure A

4.6 Dalton's law of partial pressures

pressure. First we introduce the *mole fractions* x_A and x_B. These are defined as

$$x_A = n_A/n, \qquad x_B = n_B/n \qquad (4.1.8)$$

with $n = n_A + n_B$. Then since $p_A = n_A RT/V$, $p_B = n_B RT/V$, and $RT/V = p/n$, it follows that

$$p_A = x_A p \quad \text{and} \quad p_B = x_B p. \qquad (4.1.9)$$

This dependence of the partial pressures on the composition is illustrated in Fig. 4.6. This is an exceptionally useful (and simple) way of arriving at partial pressures when the composition and total (measured) pressure of a mixture are known.

Dalton's law is another idealization that applies exactly to a perfect gas, and approximately to real gases. It captures the essential features of the behaviour of gases when they are so diffuse that their particles are completely independent.

Example: In simple applications it is often sufficient to regard the atmosphere as a mixture of 76.8% by mass of nitrogen and 23.2% by mass of oxygen. Calculate the partial pressure of each gas when the total pressure is 101 kPa (1 atm).

Method: Use eqn (4.1.9) in the form $p(N_2) = x(N_2)p$ and $p(O_2) = x(O_2)p$ where x is the mole fraction of the component. Find the mole fractions from eqn (4.1.8), but first obtain the amounts of substance of each species. Since the composition is expressed as percentages, the simplest procedure is to consider a sample of 100 g, which then contains 76.8 g nitrogen and 23.2 g oxygen. Then use $M_m(N_2) = 28.02 \text{ g mol}^{-1}$ and $M_m(O_2) = 32.00 \text{ g mol}^{-1}$ (see inside back cover).

Answer: The amounts of substance present in 100 g of air are

$$n(N_2) = \frac{76.8 \text{ g}}{28.02 \text{ g mol}^{-1}} = 2.74 \text{ mol};$$

$$n(O_2) = \frac{23.2 \text{ g}}{32.00 \text{ g mol}^{-1}} = 0.725 \text{ mol}.$$

The mole fractions of the components are therefore

$$x(N_2) = \frac{2.74}{2.74 + 0.725} = 0.791; \qquad x(O_2) = \frac{0.725}{2.74 + 0.725} = 0.209.$$

The partial pressures are therefore given by eqn (4.1.9) as

$$p(N_2) = 0.791 \times (101 \text{ kPa}) = 79.9 \text{ kPa}$$

$$p(O_2) = 0.209 \times (101 \text{ kPa}) = 21.1 \text{ kPa}.$$

Comment: A more complete description of the composition of dry air at sea level is as follows (in mass percent): $N_2(75.52)$, $O_2(23.15)$, $Ar(1.28)$, $CO_2(0.046)$, corresponding to the partial pressures (in atm) $N_2(0.782)$, $O_2(0.209)$, $Ar(0.009)$, $CO_2(0.0003)$. There are also traces of other gases, especially in towns.

Principles of physical chemistry

4.2 Accounting for the gas law

The pressure of a gas is due to the collisions its particles make with the walls of the container. Their constant battering on the walls exerts a force (the same resisting force felt when testing an inflated bicycle tyre), and there are so many collisions that it appears to us as a steady pressure (pressure is force per unit area). In this section we shall see how this picture can be used to account for the perfect gas law.

The kinetic theory

The *kinetic theory of gases* is based on the following model of a perfect gas:

(1) The gas is pictured as a collection of identical particles of mass m in continuous random motion.
(2) The particles are regarded as having zero volume (are 'pointlike').
(3) They move without interacting with each other except for the collisions that maintain their random motion.
(4) All collisions between particles and with the walls of the container are *elastic*, which means that their total translational kinetic energy after a collision is the same as that before one (similar to collisions between billiard balls but unlike those between cars).

On the basis of this model it is possible to derive the following expression for the pressure of the gas:

$$pV = \tfrac{1}{3}Nm\overline{c^2}. \tag{4.2.1}$$

V is the volume of the container, N is the number of particles present, and $\overline{c^2}$ is their *mean square speed*. (The mean square speed is the average value of the squares of the speeds of the particles.) This result is derived in the Appendix. The number of particles can be expressed as an amount of substance n through $N = nL$, L being Avogadro's constant. The product mL that then occurs in the equation is the molar mass, M_m, of the particles. Therefore the expression becomes

$$pV = \tfrac{1}{3}nM_m\overline{c^2}. \tag{4.2.2}$$

The pressure–volume relation of a perfect gas is correctly reproduced by this expression. The mean square speed of the particles is independent of the pressure and the volume (they go on hurtling around with the same average speeds when the gas is compressed so long as the temperature is the same) and so the expression has the form $pV = $ constant, in accord with Boyle's law.

The mean square speed rises when the temperature is increased and the particles move more vigorously. The precise dependence can be deduced by comparing eqn (4.2.2) with $pV = nRT$. It follows that

$$\tfrac{1}{3}nM_m\overline{c^2} = nRT, \quad \text{or} \quad \overline{c^2} = 3RT/M_m.$$

That is, the *root mean square speed* (r.m.s. speed) c_{rms} is

$$c_{rms} = \sqrt{(\overline{c^2})} = \sqrt{(3RT/M_m)}. \tag{4.2.3}$$

There are two immediate conclusions from this result. The first is that *the r.m.s. speed of particles increases as the square root of the temperature.* This means that on average the molecules in air move about 5% faster on a hot day (30 °C) than on a cold day (0 °C), and that a molecule on the surface of the sun (6000 °C) moves about $4\frac{1}{2}$ times faster on average than the same molecule in the atmosphere of the earth. The second conclusion to draw is that the *r.m.s. speed depends on the molar mass as* $1/\sqrt{M_m}$. On the average, more massive molecules move more slowly than do less massive molecules, Fig. 4.7. For instance, since $M_m = 44 \text{ g mol}^{-1}$ for carbon dioxide and $M_m = 18 \text{ g mol}^{-1}$ for water, on the average the speed of a CO_2 molecule is about $\frac{2}{3}$ that of an H_2O molecule in the atmosphere. The values of c_{rms} can be calculated very simply by substituting the values of the molar mass and the temperature into eqn (4.2.3). For carbon dioxide at 25 °C we find

$$c_{rms} = \sqrt{\left(\frac{3 \times (8.31 \text{ J K}^{-1} \text{ mol}^{-1}) \times (298 \text{ K})}{44.0 \times 10^{-3} \text{ kg mol}^{-1}}\right)} = 411 \text{ m s}^{-1},$$

or about 920 m.p.h.

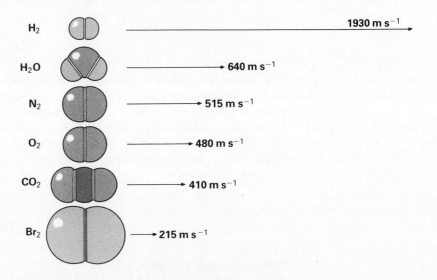

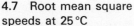

4.7 Root mean square speeds at 25 °C

The Maxwell-Boltzmann distribution

So far we have discussed only average speeds, but in a gas the particles have a range of speeds, and it is often necessary to have more detailed information. The spread of speeds is given by the *Maxwell-Boltzmann distribution*. Its predictions are sketched in Fig. 4.8. The point to note is that *the higher the temperature the greater the spread of speeds* (as well as the greater the r.m.s. speed). Although the r.m.s. speeds of small molecules (*e.g.*, O_2) at room temperature are about 500 m s^{-1}, many move much

Principles of physical chemistry

more quickly. (The ratio of the number having twice the r.m.s. speed to the number having the r.m.s. speed itself is about $1:5$.)

The spread of speeds in a gas can be measured experimentally by squirting a stream of the gas, a *molecular beam*, through a series of notched disks, Fig. 4.9. This is sometimes called the *Zartmann-Ko experiment*. The intensity of the stream that gets through is proportional to the number of molecules that have just the right speed to pass through the sequence of slits as they rotate into place. It is found that the observed distribution matches the Maxwell-Boltzmann distribution very closely, Fig. 4.10.

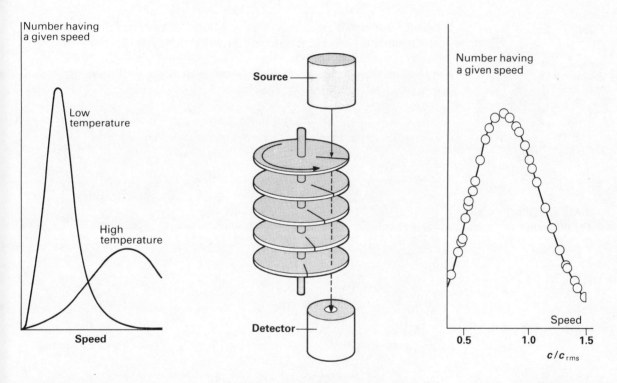

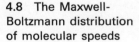

4.8 The Maxwell-Boltzmann distribution of molecular speeds

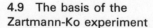

4.9 The basis of the Zartmann-Ko experiment

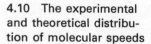

4.10 The experimental and theoretical distribution of molecular speeds

Diffusion and effusion

The process of *diffusion* is the spreading of particles from a region of high concentration to one of low concentration, Fig. 4.11(*a*), possibly through a porous barrier. Diffusion occurs in liquids as well as in gases (the spreading of sugar through an unstirred cup of tea takes place by diffusion). *Effusion* is the escape of a gas through a small hole, Fig. 4.11(*b*), as for a puncture in a tyre or a spacecraft.

The experimental observation of the rate of effusion of gases through a small hole in the side of the containing vessel led Graham (in 1829) to formulate the following law:

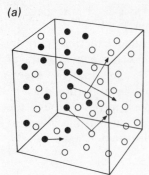

(a)

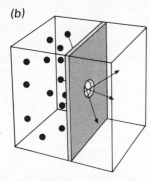

(b)

4.11 (a) Diffusion and (b) effusion

Graham's law: At constant temperature and pressure, the rates of effusion and diffusion of a gas are inversely proportional to the square root of its molar mass.

It follows that the molar masses of gases can be found by comparing their rates of effusion under the same conditions of pressure and temperature. If the time for a given volume of gas A to escape is t_A, while the time for the same volume of gas B to escape is t_B, it follows that, provided the pressure barely changes,

$$\frac{t_A}{t_B} = \frac{(\text{rate})_B}{(\text{rate})_A} = \sqrt{\left(\frac{M_{A,m}}{M_{B,m}}\right)}, \tag{4.2.4}$$

because the times are inversely proportional to the rates of escape. If the molar mass of one gas is known, the other's may be deduced.

Example: Oxygen was confined inside a vessel in the wall of which there was a small hole. It was found that 50 cm^3 of the gas escaped in 20 s. When bromine vapour was confined to the same vessel under the same conditions of temperature and pressure, the same volume escaped in 45 s. What are the molar mass and the relative molecular mass of bromine?

Method: Assume both gases behave perfectly and use Graham's law in the form of eqn (4.2.4). Take $M_r(O_2) = 32.0$; note that $M_m = M_r \, \text{g mol}^{-1}$.

Answer: From eqn (4.2.4) with A identified as Br$_2$ and B as O$_2$,

$$\sqrt{\left(\frac{M_m(\text{Br}_2)}{32.0 \, \text{g mol}^{-1}}\right)} = \frac{45 \, \text{s}}{20 \, \text{s}} = 2.25.$$

Therefore $M_m(\text{Br}_2) = (2.25)^2 \times (32.0 \, \text{g mol}^{-1}) = 162 \, \text{g mol}^{-1}$. It follows from $M_m = M_r \, \text{g mol}^{-1}$ that M_r of Br$_2$ is 162.

Comment: The precise value of M_r is 159.8. Instead of dealing with a vessel with a small hole in the wall, an easier experiment is to confine the gas in a porous pot; the same equations apply because the porosity is like a very large number of holes through which effusion can occur. Note that only a small amount of gas must be allowed to escape so that the gas pressure remains almost constant throughout the experiment.

The kinetic theory of gases easily accounts for Graham's law. The rate of effusion is proportional to the frequency with which molecules strike the area of the hole. That frequency is proportional to their r.m.s. speed, which is proportional to $1/\sqrt{M_m}$. Therefore the rate of effusion is proportional to $1/\sqrt{M_m}$, in accord with Graham's law.

Molecular diffusion through a porous barrier is the basis of one method of isotope enrichment, especially for uranium. The working material is the gas UF$_6$. The enrichment depends on ^{235}UF$_6$ (with $M_m = 349 \, \text{g mol}^{-1}$) diffusing slightly more rapidly than ^{238}UF$_6$ (with $M_m = 352 \, \text{g mol}^{-1}$). The rates of diffusion are in the ratio $\sqrt{(352/349)} = 1.004$, and so in order to achieve commercially acceptable separation the gas has to be allowed to diffuse many times through porous barriers. Simple calculations like this show why uranium separation plants are so huge (and difficult to conceal).

4.3 Real gases

In the case of a perfect gas $pV_m/RT = 1$. Measured values of pV_m/RT are plotted in Fig. 4.12. In some cases (generally for pressures below about 350 atm) the value of pV_m/RT is less than 1, while at higher presures it is greater than 1. What can be the reasons?

Molecules interact with each other, Fig. 4.13. That is an obvious conclusion from everyday experience: gases can be liquefied, their molecules sticking together on account of the attractive forces between them. (We investigate the nature of these forces in the next chapter.) Furthermore, molecules are not points; they occupy a definite volume. Therefore statement (2) defining the kinetic model is invalid, and is significantly in error when the density of the gas is so high that its molecules are close together.

The attractions between molecules have the effect of reducing the pressure of the gas. The product pV_m is therefore expected to be *less* than the perfect gas value when the attractive forces are important. That is the case at pressures up to about 350 atm. At higher pressures, when the gas is denser, the non-zero volume of the molecules becomes more important and tends to drive them apart. As a result, the product pV_m is greater than its perfect gas value. This neatly accounts for the change of pV_m/RT from values of less than 1 at moderate pressures to greater than 1 at high pressures.

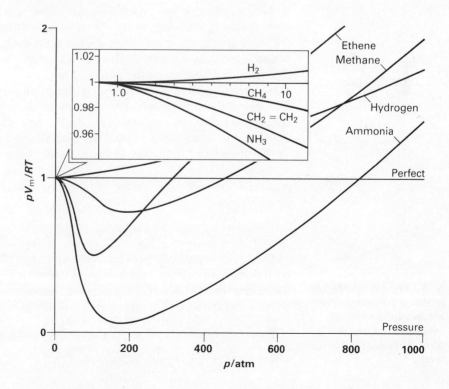

4.12 The dependence of pV_m/RT on the pressure for real gases

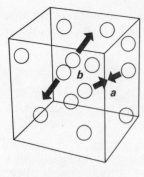

4.13 The attractive (a) and repulsive (b) terms in the van der Waals equation

Many people have tried to modify the perfect gas equation in order to obtain a better description of real gases. The first and most famous attempt was made by van der Waals (1873) who proposed the equation

$$(p + a/V_m^2)(V_m - b) = RT \tag{4.3.1}$$

in place of $pV_m = RT$. The constant a takes account of the attractive forces between particles. The constant b takes account of the repulsive forces by supposing that, as the particles themselves occupy a volume, the free volume in the container is reduced from V_m to $V_m - b$. The values a and b have been determined for many gases, and the van der Waals equation can be used to predict the properties of real gases in a handy but still approximate way.

Appendix: Calculating the pressure

Pressure is force per unit area; therefore, in order to calculate the pressure exerted by the particles inside a container of volume V, we calculate the average force they exert per unit area of the walls. For simplicity take the container to be a cube of side l (and so $V = l^3$). If the total force exerted on one of the walls is f, the pressure on the wall is force/area, or f/l^2.

The force is calculated on the basis of Newton's laws of motion. According to his second law, *the rate of change of momentum of a particle is equal to the force* acting on it. Consider, therefore, a single particle of mass m hurtling towards the wall perpendicular to the x-axis, Fig. 4.14. If its velocity along x is v_x, its momentum along x is mv_x. Immediately after the collision it has reversed its direction, and is travelling backwards with the same speed as before (the collision is elastic). Its momentum along x is now $-mv_x$. As a result of the collision with the wall there has been a change of momentum of magnitude $2mv_x$.

In order to calculate the rate of change of momentum we need to know how often the collisions occur. The molecule revisits the wall after it bounces off the one opposite (we deal below with the possibility of a collision during its journey), and the total time for the round trip is $2l/v_x$. The frequency of collisions it makes with the wall is therefore 1/(time between collisions) or $v_x/2l$. The rate of change of momentum of the particle is therefore the change of momentum on one collision multiplied by the frequency of the collisions:

Rate of change of momentum $= (2mv_x) \times (v_x/2l) = mv_x^2/l$.

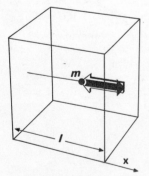

4.14 The model for calculating pressure. The arrows represent momenta before and after a collision with the wall

There are N particles present in the container. They have a wide range of different velocities backwards and forwards along the x-direction. Therefore the total rate of change of momentum of all the particles colliding with the wall is the sum of all the individual contributions:

Total rate of change of momentum $= (m/l)\{v_x^2(1) + v_x^2(2) + \cdots v_x^2(N)\}$,

where $v_x(1)$ is the velocity of particle 1 in the x-direction, and so on. The *mean square velocity* along x is the average value of v_x^2 for the entire sample:

$$\overline{v_x^2} = (1/N)\{v_x^2(1) + v_x^2(2) + \cdots + v_x^2(N)\}.$$

(The bar indicates an average value.) Therefore

Total rate of change of momentum $= (Nm/l)\overline{v_x^2}.$

This average value is valid even when particles collide in flight between the walls because on average there is always one that acquires a greater velocity when elsewhere in the sample another loses some.

The total rate of change of momentum is equal to the total force acting on the particles. But according to Newton's third law (action = reaction) this must also be the magnitude of the force experienced by the wall where the collisions are taking place. Therefore

Average force exerted on wall $= (Nm/l)\overline{v_x^2}.$

Average pressure exerted on wall $=$ force $\div$ area
$$= (Nm/l)\overline{v_x^2} \div l^2 = (Nm/l^3)\overline{v_x^2}.$$

This is the force exerted on the wall perpendicular to the x-direction. The same force is exerted on the walls perpendicular to the y and z directions, and so the *average pressure on any wall* is

$$p = (Nm/l^3)\overline{v_x^2} = (Nm/l^3)\overline{v_y^2} = (Nm/l^3)\overline{v_z^2}.$$

We now introduce the *mean square speed* of the molecules as

$$\overline{c^2} = \overline{v_x^2} + \overline{v_y^2} + \overline{v_z^2}.$$

According to the preceding expression all three contributions are the same, and so $\overline{c^2} = 3\overline{v_x^2}$. Consequently the *pressure on any wall* is

$$p = \tfrac{1}{3}(Nm/l^3)\overline{c^2}.$$

The final step is to notice that $l^3 = V$, the volume of the container. Then we arrive at

$$pV = \tfrac{1}{3}Nm\overline{c^2},$$

which is the form used in the text.

Summary

1 *Boyle's law* states that $pV =$ constant, the temperature and mass of gas being fixed.

2   *Charles's law* states that for constant pressure and mass of gas, $V \propto T$, temperature on the Kelvin scale.

3   *Avogadro's law* is that equal amounts of substance of gases occupy the same volume at the same temperature and pressure.

4   The zero of the *Kelvin scale of temperature* is the temperature at which a perfect gas occupies zero volume. The Kelvin and Celsius scales are related by $T/K = 273.15 + t/°C$.

5   The *perfect gas equation* is $pV = nRT$; real gases are described by this equation increasingly closely as the density is decreased.

6   *Standard temperature and pressure* (s.t.p.) means $0°C$ (273.15 K) and 1 atm (101.325 kPa).

7   *Dalton's law of partial pressures* states that a gas in a mixture exerts the same pressure (its *partial pressure*) as when it alone occupies the container at the same temperature.

8   The *mole fraction* of a component of a mixture is $x_A = n_A/n$, where n is the total amount of substance; the partial pressure is $p_A = x_A p$, where p is the total pressure.

9   The *kinetic theory of gases* is based on the model of a gas set out on p. 65.

10   The *root mean square speed* of molecules in a gas at a temperature T is given by $c_{rms} = \sqrt{(3RT/M_m)}$. The r.m.s. speed increases with temperature as $\sqrt{T}$ and decreases with molar mass as $1/\sqrt{M_m}$.

11   The *Maxwell-Boltzmann distribution* gives the proportion of particles in a gas having a given speed. The higher the temperature the greater the spread of speeds in the sample, and the greater the r.m.s. speed.

12   *Graham's law* states that at constant temperature and pressure the rates of effusion and diffusion of a gas are proportional to $1/\sqrt{M_m}$.

13   The value of pV_m/RT for real gases is generally less than 1 at low pressures (from zero to around 350 atm) on account of the *attractive interactions* between molecules; at higher pressures it is greater than 1 on account of the *repulsive effects*.

14   The *van der Waals equation* $(p + a/V_m^2)(V_m - b) = RT$ is an approximate description of real gases; a represents the effects of attractive interactions between molecules and b represents the effects of their repulsive interactions.

Problems

1   Outline the experiments that led to the formulation of the perfect gas law.

2   Explain non-mathematically why (*a*) a gas fills any container it occupies, (*b*) a gas exerts a pressure on the walls, (*c*) the pressure of a given amount of gas increases when its temperature is increased in a container of constant volume.

3   A sample of gas supports a column of mercury 720 mm high in an open-ended manometer; what is the pressure of the gas? Express the answer in atm and kPa. (Density of mercury, $13.6 \, g \, cm^{-3}$; acceleration due to gravity, $9.81 \, m \, s^{-2}$; 1 atm corresponds to 101.3 kPa.)

4   The cathode ray tube of a television set has a volume of about $5 \, dm^3$. The pressure inside it is typically 0.10 Pa at 25 °C. Estimate the number of molecules of air it contains.

5   When the set in Question 4 is operating the temperature inside the tube rises to 80 °C. What is the pressure?

6   A typical plant used for the production of ammonia by the Haber-Bosch process stores the ammonia at 720 K and 3×10^7 Pa before transferring it to a cooling plant for storage as liquid. On the assumption that the gas behaves perfectly under these conditions, estimate the amount of substance of NH_3 that the $100 \, m^3$ vessel stores. What mass of ammonia does that represent?

7   A balloon of volume $1000 \, m^3$ was filled with helium at 27 °C to a pressure of 1.01 atm. Upon ascending the temperature fell to -10 °C while the external pressure fell to 0.740 atm. What mass of helium must be released in order to keep the volume constant?

8   A vessel of volume $5 \, dm^3$ was filled with (a) $2 \, mol \, H_2$, (b) then a further $1 \, mol$ Cl_2, and then (c) the mixture was sparked, leading to the formation of $2 \, mol$ HCl (g). What are the total pressure and the partial pressures of the components at each stage, the temperature being 25 °C throughout?

9   Calculate the root mean square speeds of (a) Cl_2, (b) CH_4 molecules at s.t.p. What influence on the speeds does an increase of pressure at constant temperature have? What is the ratio of the speeds of $^{37}Cl_2$ to $^{35}Cl_2$?

10   Observation of the spectra of ^{57}Fe atoms on the surface of the sun shows (from the Doppler shift) that they have a root mean square speed of about $1.6 \, km \, s^{-1}$. What is the temperature of the sun's surface?

11   A small volume of gas of unknown molar mass was allowed to effuse from a container and compared with the time that it took the same volume of hydrogen chloride to effuse under the same conditions. While hydrogen chloride took $90 \, s$ to effuse, the other gas required only $60 \, s$. What is its molar mass?

12   List the assumptions on which the kinetic theory of gases is based. Which assumptions are false, and what effects do they have on real gases?

13   The molar volume of carbon dioxide was found to be $1.32 \, dm^3 \, mol^{-1}$ at 48 °C and 18.4 atm pressure. Compare this observation with the pressure calculated on the basis of (a) the perfect gas equation, (b) the van der Waals equation (the values of the constants for carbon dioxide are $a = 3.59 \, dm^6 \, atm \, mol^{-2}$ and $b = 4.27 \times 10^{-2} \, dm^3 \, mol^{-1}$).

14   (a) List the main assumptions of the kinetic theory for ideal [perfect] gases.
    (b) Give the origins of two of the main differences between the behaviour of real and ideal gases.
    (c) State the dependence of the root mean square velocity of a gas on (i) the relative molecular mass, (ii) the temperature.
    (d) Calculate the root mean square velocity of [nitrogen] (N_2) molecules at 7 °C. [$A_r(N) = 14$, $R = 8.31 \, J \, mol^{-1} \, K^{-1}$.]

(Welsh part question)

15   Discuss briefly, but critically, the basic postulates underlying the kinetic theory of gases and explain

    (a) gas pressure,
    (b) thermal expansion at constant pressure,
    (c) the use of $\overline{c^2}$, the mean square velocity, in the equation

$$pV = \tfrac{1}{3}mN\overline{c^2}$$

where N molecules of ideal [perfect] gas each of mass m, occupy a volume V at a pressure p.

Calculate:

- (*i*) the kinetic energy (in joules) of the molecules in one mole of ideal gas at 47 °C;
- (*ii*) the *root* mean square velocity of hydrogen iodide molecules, HI, at 47 °C in the gaseous phase (in this calculation the molar mass of hydrogen iodide must be expressed in the appropriate SI unit, *i.e.*, the *kilogram*).
- (*iii*) the *ratio* of the *root* mean square velocities of oxygen, O_2, and hydrogen iodide at 47 °C;
- (*iv*) the time expected to be taken for a given volume of hydrogen iodide at 47 °C to effuse (or diffuse) through a pin-hole if the same volume of oxygen under the same conditions takes 60 s.
- (*v*) Compare this with that predicted by Graham's law of diffusion, which should be stated.

(SUJB)

16 In the kinetic theory of gases, what basic assumptions are made in the derivation of the expression $pV = \frac{1}{3}Nm\overline{c^2}$ for an ideal [perfect] gas?

Use this expression (*a*) to derive Graham's law of diffusion, (*b*) to calculate the root mean square speed $(\sqrt{\overline{c^2}})$ for argon at s.t.p., and (*c*) to calculate the kinetic energy for one mole of an ideal gas in terms of the gas constant, *R*, and the temperature, *T*.

Sketch the curve showing the distribution of molecular speeds in a gas and mark on the curve the position of the *average speed*, the *most probable speed* and the *root mean square speed*.

Describe how the distribution of molecular speeds changes when the temperature is increased.

Explain how the gas equation $pV = nRT$ has been modified to take account of the behaviour of *real* gases.

(Oxford and Cambridge)

17 Derive an expression relating the pressure and volume of a gas to the root mean square speed of the molecules, c_{rms}. Write down a formula for the number of molecules striking unit area of the wall per second, taking the mean speed $\bar{c}$ to be equal to $0.92c_{rms}$. Suppose a container is filled with a mixture of two gases A and B with relative molecular masses (molecular weights) M_A and M_B, and that there is a small orifice in the wall through which they escape or effuse. Write down an expression for the relative rates of effusion of A and B, stating any assumptions made in the derivation. The element chlorine has two stable isotopes ^{35}Cl and ^{37}Cl. What would be the relative rates of effusion of chlorine molecules through a small orifice?

(Oxford Entrance)

18 State the ideal gas laws. Show how they must be modified to explain the behaviour of real gases.

Two glass bulbs of capacity $0.2\,dm^3$ and $0.1\,dm^3$ joined only by a narrow capillary contain air at 288 K and 1 atmosphere pressure. Find the pressure of the air when the larger bulb is heated by steam to 373 K whilst the smaller remains at 288 K. State any simplifying assumptions you may make.

(Oxford Entrance)

5   Inside solids

In this chapter we look at the structures of solids. We examine the forces that hold atoms, ions, and molecules together, and look at ways of discovering their arrangement inside solids by X-ray diffraction. Then we turn to crystals and see something of the wide range of structures that are responsible for most of what is permanent in the world around us.

Introduction

Matter often comes in lumps. In contrast to gases, where the particles are dispersed almost completely chaotically and are in constant motion, the particles inside solids are almost stationary and located in ordered arrays.

The importance of solids in everyday life hardly needs to be stressed. Much of modern technology is concerned with the modification of the structures of solids, either in their external form, as in the manufacture of cars out of ingots of steel, or in the alteration of their internal structures in order to achieve special properties. Examples include the development of steels with improved corrosion resistance and greater strength, the development of new semiconductors, and the manufacture and application of synthetic polymers.

5.1   The forces of aggregation

There are forces that cause particles to stick together even when the valencies of their atoms have been fully satisfied. The clearest evidence for this is the existence of *condensed phases* (*i.e.*, liquid and solid forms) of the noble gases. At low enough temperatures these forces can overcome the chaotic thermal motion of the atoms and cause them to congregate into liquids and solids, Fig. 5.1.

Covalent structures

Some lumps of matter are single huge molecules. In them the atoms are linked together into extensive covalently bonded structures. A famous example is *diamond*, where every carbon atom forms a bond with each of four neighbours at the points of a regular tetrahedron, Fig. 5.2. This structure is repeated throughout the crystal, and so a 1 carat diamond (of

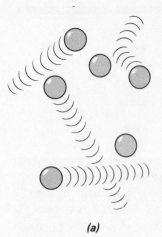

(a)   (b)   (c)   (d)

5.1 Lowering the temperature of a sample of (*a*) a gas leads to condensation to (*b*) a liquid which freezes to (*c*) a solid, which is totally free of vibrational excitation at (*d*) absolute zero

mass 200 mg) can be thought of as a single molecule of about 1.0×10^{22} atoms. Since it is difficult to break the strong C—C bonds, and because the structure is like the steel framework of a large building, the diamond crystal is hard and rigid. The *graphite* structure is quite different, Fig. 5.3. It consists of planar sheets of carbon atoms arranged in hexagons (like chicken wire). The sheets are only weakly bound together, and slide over each other easily. That accounts for graphite's slipperiness. The rigidity is typical of other covalently bonded solids (such as quartz, where the crystal is a network of SiO_2 units).

Metals

We have seen that a metal is an orderly collection of ions surrounded by and held together by a sea of electrons, Fig. 5.4. The diagram shows that

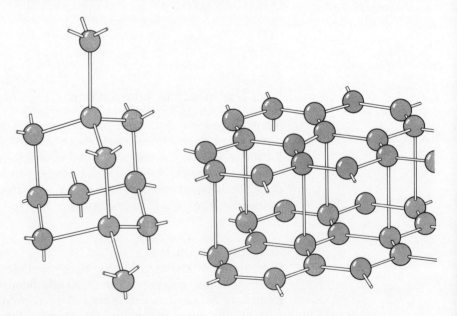

5.2 The structure of diamond   5.3 The structure of graphite

Principles of physical chemistry

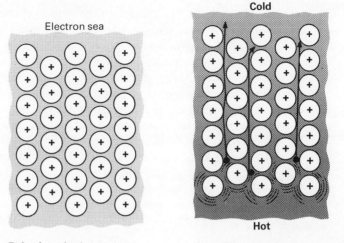

Electron sea

Cold

Hot

5.4 A univalent metal   5.5 The conduction of heat by electrons

the bonding does not have the directional qualities of a covalent solid like diamond, and so the sea of electrons allows the groups of ions to be pushed into a new arrangement. The contrast between the directional nature of a covalent bond and the non-directional nature of a metallic bond is the reason why it is far easier to bend a piece of metal than to bend a diamond or a crystal of quartz. Shifting groups of ions through the electron sea can be brought about in a number of ways. These include hammering, rolling, extruding, stretching, and bending. These qualities of *ductility* (the ease of drawing out) and *malleability* (the ability to be formed by hammering) are typical of metals.

The presence of the sea of mobile electrons also accounts for the other typical characteristics of metals: their *electrical and thermal conductivities* and their ability to reflect light. Electrical conductivity arises from the ability of the electrons to flow through the ion framework when there is a potential difference applied across the ends of the metal (ends that may be miles apart if the metal is a part of a national grid). Thermal conductivity arises in a similar way. The electrons can pick up kinetic energy from the vibrations of the ions where the metal is hot, pass rapidly through the ion framework until they collide with a distant ion and so cause it to vibrate more vigorously, Fig. 5.5. In other words, the mobile electrons are efficient transporters of energy. The reflecting power of metals is another aspect of the mobility of the electrons; an incident light wave forces the electrons near the surface of a metal to oscillate, and as a result the incident light is reflected back out instead of penetrating into the bulk. Our reflection in a mirror is the oscillation of the silver's mobile electron sea.

Ionic structures

Common salt is not a gas that has to be bubbled through food. Even though the sodium and chlorine valencies are satisfied (in the sense of having closed-shell configurations), sodium chloride is a rigid solid. The reason lies in the electrostatic *Coulomb forces* arising from the charges on the two kinds of ion. These forces act in all directions, and so it is

Inside solids

energetically favourable for other ions to cluster near the first pair, and then for others to cluster near them. The usual form of sodium chloride is therefore an extensive aggregate of positive and negative ions held together by their mutual electrostatic attraction. This is the common feature of all ionic species, such as the other alkali halides, calcium carbonate, magnesium sulphate, and so on.

A *crystal lattice* is a three-dimensional orderly arrangement of ions. The strengths of ionic crystal lattices are due to the strengths of the electrostatic interactions between the ions, and only when the crystal has been heated to a high temperature (801 °C in the case of sodium chloride) do the ions shake around so violently that the electrostatic forces are overcome and the crystal melts into a liquid. In Chapter 9 we look at ways of measuring the strengths with which ionic lattices are bound.

Van der Waals forces

We now turn our attention to solids formed from discrete, non-ionic particles. When the particles are molecules these are called, *molecular solids*. How, for instance, do molecules of methane or of benzene stick together, and what forces are responsible for the condensation of the noble gases?

The *van der Waals forces* between discrete covalent molecules are due to interactions between dipoles. There are two principal types, the *dipole–dipole forces* and the *dispersion forces* (or *London forces*).

Dipole–dipole forces can occur only when both molecules are polar (have permanent electric dipoles). From Fig. 5.6 it can be seen that when the molecules are in the orientations shown, there is a favourable electrostatic interaction between neighbouring opposite charges. As a result they stick together.

Dispersion forces occur between polar molecules and between nonpolar molecules, and their effect normally dominates any dipole–dipole forces that are also present. They occur even between atoms. Their origin is found in the fact that the electron distribution of a molecule should not be thought of as being static and frozen, but as continuously flickering from one arrangement to another. Two nearby molecules interact through their *instantaneous* dipoles, Fig. 5.7, and the interaction is strong enough to account for the condensation of the noble gases, the hydrocarbons, iodine, and so on.

Van der Waals interactions are much weaker than ionic forces, and so molecular solids are often soft. Margarine, for instance, is a collection of molecules sticking together through van der Waals interactions, and a knife blade can readily reorganize the arrangement. The boiling and melting temperatures of van der Waals materials are also low (in comparison with ionic crystals) because little thermal agitation is needed to overcome the interactions. Van der Waals interactions are largest between large molecules because dispersion forces are largest when the molecule's electron clouds are extensive (the nuclei then have only a weak grip on the outer electrons, and so the instantaneous dipole moment can flicker between large values).

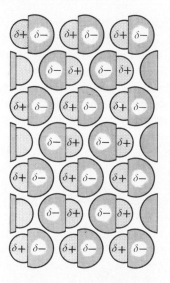

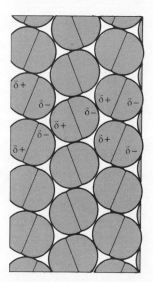

5.6 Dipole–dipole forces and the structure of solid hydrogen chloride

5.7 Dispersion forces and the structure of solid iodine

Example: The boiling temperatures of three of the hydrogen halides under 1 atm pressure are HCl: 188 K; HBr: 206 K; HI: 238 K. Account for the trend in values.

Method: Note that all three compounds are discrete, covalent molecules in both the liquid and the vapour. The intermolecular forces that must be overcome are dipole–dipole forces, dispersion forces, and hydrogen bonding (see below). Think about the permanent dipole moments on the basis of the electronegativities of the halogen atoms involved (Table 2.1).

Answer: The electronegativities of the halogen atoms increase along the series I, Br, Cl, and therefore the magnitudes of the permanent dipole moments of the molecules increase along the series HI, HBr, HCl. On the basis of dipole-dipole interactions we expect the boiling temperatures to increase in the same order. This is contrary to observation. Dispersion forces are largest between large, many-electron species. The number of electrons increases along the series HCl, HBr, HI, and so on the basis of dispersion forces we expect the boiling temperatures to increase in the same order; this is in agreement with the data. We therefore conclude that the dominant contribution to the intermolecular forces in these compounds is from the dispersion forces.

Comment: This *Example* reinforces the point made in the text that dipole–dipole forces rarely exert a dominating influence on boiling temperatures.

Inside solids

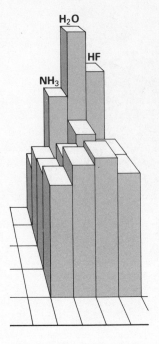

H₂O

HF

NH₃

5.8 The boiling temperatures of some hydrides

Intermediate in structure between molecular solids and covalent solids are many of the polymeric materials that are such a characteristic feature of modern life. As an example, consider polythene, poly(ethene). This consists of long chains of covalently bonded —CH_2CH_2— units. The chains themselves stick together as a result of their van der Waals interactions. Modern plastics also contain additives and covalent cross-links between the chains.

Hydrogen bonds

A hydrogen bond is a special kind of bond in which the link is due to a proton lying between two very electronegative atoms, such as oxygen, fluorine, or nitrogen. We have seen that the O—H bond is polar in the sense $O^{\delta-}$—$H^{\delta+}$, and the electron withdrawing effect of oxygen is so great that the proton is partially exposed and can attach itself electrostatically to some other region of high electron density, such as the lone pairs of oxygen or nitrogen atoms of other molecules. The bond is often denoted $X^{\delta-}$—$H^{\delta+}$---$Y^{\delta-}$, where X and Y are strongly electronegative atoms.

Hydrogen bonding is of great importance for the structure of the condensed phases of water. Its role in the liquid can be seen by examining the boiling temperatures of the hydrides of some elements, Fig. 5.8: liquid water has an anomalously high value (so too have ammonia and hydrogen fluoride). The explanation is that the hydrogen bonds bind the molecules into a liquid instead of leaving them as a gas. As a result, water is available in the world as a liquid and not as a gas, and life has been able to evolve.

The hydrogen bonds of water remain when it solidifies. They then hold the molecules into an open crystal structure, Fig. 5.9. The openness of this structure is also important for the existence of life. This is because the hydrogen bonds not only help to hold the water molecules rigidly as a solid, but also hold the molecules apart, so that ice is less dense than water. As a result, the tops of ponds and oceans freeze first and help to insulate the lower regions, and aquatic life is not inhibited.

Hydrogen bonding often survives even in the gas phase. For example, ethanoic acid vapour contains a high proportion of dimers, Fig. 5.10, and this affects its apparent molar mass. In liquid hydrogen fluoride the hydrogen bonding is so strong that the molecules stick together into

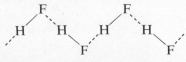

clusters, which survive partially in the vapour. A number of important solids have structures determined largely by hydrogen bonds. One example is sugar, built from molecules (sucrose) bristling with oxygen and hydrogen atoms. Another is wood, which consists of long chains of cellulose and gets its strength from the hydrogen bonds between them. Hydrogen bonds also play a role in proteins because they organize their long flexible polypeptide chains into definite arrangements, and in biology the shape of a molecule plays an important role in its function.

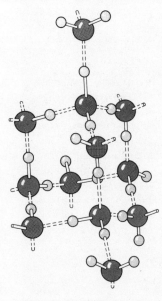

5.9 The structure of ice

Principles of physical chemistry

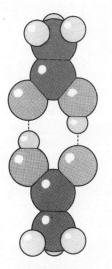

5.10 The hydrogen-bonded dimer of ethanoic acid

5.2 Investigating crystals

The information on the structure of the solids that we have been describing has come from experimental observations. The most helpful technique has been a version of X-ray diffraction (Section 3.3). X-ray diffraction was first used to discover the way that particles are stacked together to form simple crystals, and in this application is known as *X-ray crystallography*, Box 5.1.

Box 5.1: The development of X-ray crystallography

The origin of modern X-ray crystallography appears to have been a remark made by Max von Laue concerning the then recently discovered X-rays. He put forward the view (in 1912) that, since radiation is diffracted by objects of a size similar to its wavelength, X-rays (which were thought to have wavelengths of the order of 100 pm) ought to undergo diffraction by the parallel planes of ions in a crystal, which are spaced by approximately that distance. The remark was taken up both by von Laue himself and his collaborators Friedrich and Knipping, and also by the Braggs, William and his son Lawrence, who later were to receive the Nobel prize for their joint work. Initially the technique was used to obtain the dimensions of crystals and to establish the arrangement of ions in simple compounds. Immense progress came with the introduction of electronic computers, which are able to handle the vast amount of data that comes from the analysis of single crystals of substances diffracting in all directions, and to deal with the complicated and extensive calculations that have to be done in order to extract the detailed information lying within the data. The gathering of data is now extensively automated, and extremely complex molecules can be investigated. Among the landmarks of the applications of the technique are the determination of the structures of penicillin and vitamin B_{12} (by Dorothy Hodgkin in 1949 and 1955, respectively), of lysozyme (by D. C. Phillips in 1965), of haemoglobin (by J. C. Kendrew and M. Perutz in 1955), and the most famous biological molecule of all, DNA (by M. H. F. Wilkins, F. H. Crick, and J. D. Watson in 1953). All these structures took many years of work, even with computers, yet the enormous advances that have taken place in molecular biology in the last decade are a direct consequence of the information they provide.

X-ray crystallography

The basic idea behind the application of X-rays to the determination of the structures of crystals is the interference that occurs when the rays are 'reflected' by the layers of atoms. This is illustrated in Fig. 5.11.

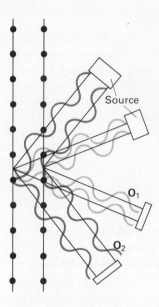

The rows of dots represent the centres of the particles of the crystal, and the observer at O_1 sees waves that have been reflected by all the layers (only two are shown in the illustration). Although the two rays have travelled different distances, their crests or troughs coincide at O_1, and so the observer detects X-rays there. An observer at O_2, on the other hand, receives waves that are exactly out of step ('out of phase'); the crests cancel the troughs, and the intensity is zero. If the spacing between the planes is different from that shown, the detector has to be moved to a different place in order to give a signal. The setting of the detector therefore depends on the spacing between the layers, which can therefore be measured.

The basic equation of X-ray diffraction is the *Bragg equation*, which relates the angle θ (Fig. 5.12) at which intensity can be observed to the spacing, d, between the planes:

Bragg equation: $\lambda = 2d \sin \theta.$ (5.2.1)

λ is the wavelength of the X-rays, and is typically about 100 pm.

5.11 X-ray diffraction by a crystal lattice

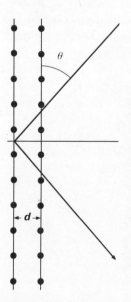

5.12 The geometry for the Bragg equation (d, layer spacing; θ, glancing angle)

Example: When a sodium chloride crystal is investigated with X-rays of wavelength 154 pm (generated by electron bombardment of a copper target) a strong diffraction intensity is observed at $\theta = 15.9°$. What is the separation between neighbouring planes of ions?

Method: Use the Bragg equation, eqn (5.2.1), to find the plane separation, d.

Answer: Eqn (5.2.1) rearranges to

$$d = \frac{\lambda}{2 \sin \theta} = \frac{154 \times 10^{-12} \text{ m}}{2 \sin 15.9°} = 2.81 \times 10^{-10} \text{ m}.$$

This distance corresponds to 281 pm since $1 \text{ pm} = 10^{-12} \text{ m}$.

Comment: It is possible to draw many different types of planes through the ions in a crystal, and so in a practical analysis of a diffraction pattern we have to decide which planes we are investigating before their separations can be interpreted in terms of the crystal structure. In modern work this kind of analysis is virtually fully automated and carried out on a computer.

In practical applications of X-ray diffraction the intensities are recorded all over a sphere surrounding the crystal, and the crystal itself is rotated. The pattern of intensities then gives very detailed information about the spacing of all the layers. This is the basis of modern knowledge about the interiors of solids.

Principles of physical chemistry

The natures of metallic crystals

The position of atoms or ions inside a crystal is called the *crystal lattice*. The description of crystals is simplified by deciding on the size and shape of the *unit cell*. This is the fundamental block from which, by stacking them together, the entire crystal can be constructed. It turns out that there are only 14 possible shapes (but they can come in any size). Three of the simplest are illustrated in Fig. 5.13.

The simplest crystal structures are obtained when all the particles (*e.g.*, ions) are the same. This is the case with metals. Every positive ion can be regarded as a small sphere, and the crystal structure is the result of stacking them together in large numbers in a regular pattern.

The way this works in practice can be seen by considering Fig. 5.14, which shows several layers of *close-packed spheres*. The three-dimensional structure is built up by placing layers one on top of each other. The spheres of the second layer lie in the dips of the bottom layer. The spheres of the third layer lie directly over the spheres of the bottom layer. This gives a *close-packed structure* known as *hexagonal close packing* (h.c.p.). If the bottom layer is denoted A and the next B, then, since the third layer repeats the first, h.c.p. can be expressed as ABABA ... This is the structure shown by magnesium and zinc, among others. (Solid helium is also h.c.p.)

The alternative placing of the third layer is over positions occupied by neither A nor B, Fig. 5.15. The structure is denoted ABCABC ... and is called *cubic close packing* (c.c.p.). Among the metals having this structure are aluminium, copper, silver, and gold. (Solid argon and neon are also both c.c.p.)

Careful inspection of the two close-packed structures shows that each sphere has 12 immediate neighbours. This number is the *coordination number* of the lattice. Some metals have less densely packed structures.

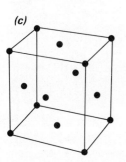

(a)

(b)

(c)

5.13 The (a) simple cubic, (b) body-centred cubic, and (c) face-centred cubic unit cells

5.14 Hexagonal close packing

5.15 Cubic close packing

For instance, the ions of potassium and iron (in one of its forms) lie at the positions corresponding to a *body-centred cubic* (b.c.c.) lattice (Fig. 5.13). This has a coordination number of 8. Only one metal, polonium, is known to have the *primitive cubic structure*, a cube with an ion at each corner, Fig. 5.13. Its coordination number is 6, and the structure is only loosely packed.

The natures of ionic crystals

While metals are structures built from identical particles, *ionic crystals* have at least two kinds, and they have opposite charges. As a result, their structures are more complicated. Since opposite charges attract each other there is a tendency for ions of one charge to group round ions of the other charge. A further complication is that different ions have different sizes, and so the structure has to take into account the geometrical problem of stacking together spheres of different sizes. The problem is rather like the best way of packing oranges and grapefruit together in the same box.

The relative sizes of some ions are illustrated in Fig. 5.16. Positive ions tend to be smaller than their parent atoms because the same nuclear charge attracts fewer electrons. Negative ions tend to be larger than their parent atoms because more electrons are under the control of the same nuclear charge.

The solutions that nature finds to the problems of stacking ions of different size together, so as to arrive at a structure that is both electrically neutral overall and has the lowest energy, can be illustrated by the structures of sodium chloride and caesium chloride. In the case of sodium chloride the ratio of the radii of the ions, the *radius ratio* (the ratio [radius smaller]/[radius bigger]) is 0.564, the radius of Na^+ being only about half that of Cl^-. In contrast the radius of Cs^+, being lower down Group I, is

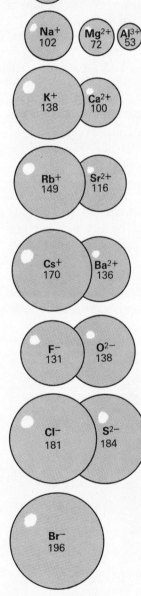

5.16 Ionic radii (in picometres, 10^{-12} m)

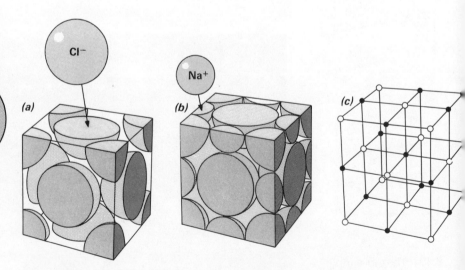

5.17 The sodium chloride structure: (a) the f.c.c. Cl^- lattice, (b) the Na^+ ions enter the holes, (c) the resulting two interpenetrating f.c.c. lattices

Principles of physical chemistry

similar to that of Cl^- and the radius ratio is 0.939. X-ray crystallography shows that, although the ions are of the same charge type, the crystals have markedly different structures.

Figure 5.17 shows the structure adopted by sodium chloride (*rock salt*). Concentrate first on the arrangement of the chloride ions. They are arranged at the corners of a cube, and there are others at the centres of each face. This structure is *face-centred cubic* (f.c.c.), Fig. 5.13. The Cl^- lattice is open and the ions are not touching (that would be energetically very unfavourable). The Na^+ ions are also arranged in an open f.c.c. structure. This lattice fits comfortably into the gaps of the Cl^- lattice, and so the structure consists of *two interpenetrating f.c.c.* lattices. In the complete structure each Na^+ ion is surrounded by six nearest neighbour Cl^- ions, and vice versa, and so rock salt has 6,6-*coordination*.

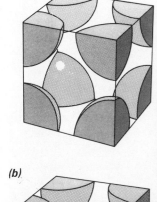

(a)

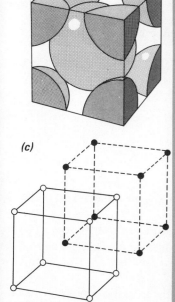

(b)

(c)

5.18 The caesium chloride structure: (a) the simple cubic Cl^- lattice, (b) a Cs^+ ion enters the hole, (c) the resulting two interpenetrating simple cubic lattices

Example: The density of sodium chloride is 2.163 g cm^{-3} and the size of its unit cell is 564.1 pm. Deduce a value for Avogadro's constant.

Method: Calculate the density of the crystal on the basis of molar mass and the volume of the unit cell, and then compare that density with the measured value.

Answer: Refer to Fig. 5.17. Each unit cell contains 4 Na^+ ions and 4 Cl^- ions. This is arrived at by noting that the 8 Cl^- ions at the corners of the cell each project only one octant (one eighth of a sphere) into the cell, and so contribute one Cl^- overall, while the 6 Cl^- on the faces each project one hemisphere into it, and so contribute 3 Cl^- overall, and hence there are 4 Cl^- in total. A similar calculation accounts for the 4 Na^+. As the side of the unit cell is 564.1 pm the volume occupied by these 4 NaCl units is $(564.1 \times 10^{-12} \text{ m})^3 = 1.795 \times 10^{-28} \text{ m}^3$. Therefore the volume occupied by unit amount (*i.e.*, 1 mol NaCl) is $\frac{1}{4} \times L \times (1.795 \times 10^{-28} \text{ m}^3)$. The mass of unit amount of NaCl is the molar mass, 58.44 g mol^{-1}. Therefore the density is the ratio

$$\frac{\text{molar mass}}{\text{molar volume}} = \frac{58.44 \text{ g mol}^{-1}}{\frac{1}{4} \times L \times (1.795 \times 10^{-28} \text{ m}^3)}.$$

Since we also know that the density is 2.163 g cm^{-3} ($= 2.163 \times 10^6 \text{ g m}^{-3}$) the two expressions can be equated and solved for L:

$$L = \frac{58.44 \text{ g mol}^{-1}}{\frac{1}{4} \times (2.163 \times 10^6 \text{ g m}^{-3}) \times (1.795 \times 10^{-28} \text{ m}^3)} = 6.021 \times 10^{23} \text{ mol}^{-1}.$$

Comment: The accepted value of Avogadro's constant is $6.022 \times 10^{23} \text{ mol}^{-1}$.

Now turn to the caesium chloride structure. X-ray diffraction shows that it is different from the sodium chloride structure, the basic reason being that the Cs^+ ions are too big to fit into the gaps in the Cl^- f.c.c. lattice. Figure 5.18 shows how the lattice is assembled. The Cl^- are at the points of a primitive cubic lattice (Fig. 5.13). The Cs^+ lattice is also built from primitive cubic unit cells. There is a big central empty region in each Cl^- unit cell, and the Cs^+ ion fits into it. Therefore the lattice consists of two interpenetrating primitive cubic lattices. Each Cs^+ ion is surrounded by eight Cl^- ions, and vice versa, and so the lattice has 8,8-*coordination*.

The radius ratio gives an important clue to the likely nature of the crystal structure of ionic materials. It follows from geometrical considerations that 8,8-coordination should be expected when the radius ratio exceeds $\sqrt{3}-1 = 0.732$; for a smaller ratio 6,6-coordination is expected.

Example: Predict the likely crystal structure of potassium bromide.

Method: Make the prediction on the basis of the radius ratio r(smaller)/r(bigger). If the ratio is less than 0.732 the likely structure is the rock salt structure; if it exceeds 0.732 the likely structure is the caesium chloride lattice. Ionic radii are given in Fig. 5.16.

Answer: The radius ratio is

$$\frac{r\text{(smaller)}}{r\text{(bigger)}} = \frac{r(K^+)}{r(Br^-)} = \frac{138\,\text{pm}}{196\,\text{pm}} = 0.704.$$

Since the radius ratio is less than 0.732 we predict that potassium bromide should have a rock salt structure.

Comment: X-ray diffraction studies confirm this prediction. Sometimes, though, the radius-ratio prediction fails. In the case of potassium fluoride the radius ratio is 0.949 yet it has the rock salt structure.

A summary of the wide variety of solid forms, and some of their typical properties, is given in Table 5.1.

Summary

1 *Covalent structures* are essentially single large molecules. The structures are hard and rigid.
2 A *metal* is an orderly array of positive ions in a sea of electrons. The bonds are *non-directional*, and so metals are *malleable* and *ductile*. The free electrons account for the *thermal and electrical conductivities* of metals and for their *reflecting power*.
3 *Ionic structures* are held together by the electrostatic attraction between oppositely charged ions.
4 A *crystal lattice* is the three-dimensional arrangement of ions (or atoms) in space.
5 *Van der Waals forces* are the forces that bind covalent molecules into a condensed phase: they are classified as *dipole–dipole forces* and *dispersion forces*.
6 The *dipole–dipole forces* are the forces between the permanent electric dipoles of polar molecules.

Table 5.1 Bonding and the nature of solids

	Simple molecular solids	Ionic crystals	Metals	Giant molecules	Polymeric
Constituents	molecules	ions	ions and electrons	atoms	molecules
Bonding	covalent within molecules, van der Waals and H-bonding between molecules	coulombic	metallic	covalent between atoms	covalent within molecules, van der Waals between molecules (and entanglement)
Examples	sucrose, I_2	NaCl, CaO, Na_2SO_4, $CaCO_3$	Cu, Na	SiO_2, C, SiC	poly(ethene), rubber
Melting temperature	low	high	high	v. high	moderate
Boiling temperature	low	high	high	v. high	often decomposed
Physical nature	soft	hard but brittle	ductile, malleable	hard	plastic
Usual solubility:					
(1) Water	insoluble	soluble	insoluble	insoluble	insoluble
(2) CCl_4	often soluble	insoluble	insoluble	insoluble	insoluble

7 The *dispersion forces*, or *London forces*, arise from the interaction of transient dipoles in the charge distributions of molecules. They occur between all types of molecules (and atoms).

8 Solids bound together by van der Waals forces tend to be soft and to have low melting and boiling temperatures.

9 A *hydrogen bond* is the link $X^{\delta-}$—$H^{\delta+} \cdots Y^{\delta-}$ where X and Y are strongly electronegative atoms. An important role for hydrogen bonds is in the structure of ice and in the shapes adopted by biological molecules such as proteins.

10 The principal method for the determination of the structures of crystals is *X-ray crystallography*, and the basic equation is the *Bragg equation*, eqn (5.2.1).

11 The *unit cell* of a crystal is the basic structure which, if stacked together, reproduces the structure of the crystal lattice. There are only 14 types of unit cell.

12 The structures of *metals* can be explained in terms of the geometrical problem of packing together identical spheres. Two *close-packed* structures are possible.

13 One close-packed structure is *hexagonal close packing* (h.c.p.). This is an ABABA... collection of close-packed layers; the *coordination number* (the number of nearest neighbours) is 12.

14 The other close-packed structure is *cubic close packing* (c.c.p.). This is an ABCABC... collection of close-packed layers; the coordination number is also 12.

15 The *body-centred cubic* (b.c.c.) structure has coordination number 8 and is less closely packed.

16   The structures adopted by ionic crystals depend on the *radius ratios* *r*(smaller)/*r*(bigger) of the ions. The alkali halides have the sodium chloride or *rock salt structure* (two interpenetrating f.c.c. lattices) when the radius ratio is less than 0.732 or the caesium chloride structure (two interpenetrating primitive cubic lattices) if the radius ratio lies between 0.732 and 1.000.

17   The sodium chloride lattice has 6,6-*coordination*; the caesium chloride lattice has 8,8-*coordination*.

Problems

1   List the properties characteristic of (*a*) ionic materials, (*b*) materials formed from individual covalent molecules, (*c*) giant covalent structures, (*d*) metals.

2   Why does copper expand when it is heated? Why does ice contract when it melts?

3   Why are metals (*a*) good conductors of electricity, (*b*) good conductors of heat, (*c*) highly reflective? Explain why some metals are coloured (*e.g.*, yellow gold and pink copper).

4   Describe the structures of (*a*) diamond and (*b*) graphite, and list and explain their different physical properties.

5   Describe the structure of the hydrogen bond. State when it is likely to occur, and show how its presence affects the physical properties of (*a*) water, (*b*) ethanoic acid, (*c*) sugar, (*d*) wood. Why is it easier to chop and saw wood along the grain rather than across?

6   Classify into types of solid the following materials: (*a*) table salt, (*b*) wax, (*c*) hair, (*d*) wood, (*e*) detergents. (*f*) soap, (*g*) bricks.

7   Explain how the close-packing of identical spheres can lead to either a cubic close-packed or a hexagonal close-packed structure. Give two examples of each type of structure.

8   Explain why the following statements are wrong: (*a*) the rock salt (sodium chloride) structure is face-centred cubic; (*b*) the metal polonium, which has a primitive cubic unit cell with the atoms arranged at the corners of a cube, has a close-packed structure.

9   State the different types of van der Waals interactions that may occur between covalent molecules. Which is usually the more important?

10   Explain the term *radius ratio* and its significance. Calculate the ratio for the alkali bromides on the basis of the information in Fig. 5.16 and predict the form of the crystal structure in each case.

11   Describe the forces between the constituent species in solids considering the relative magnitude of each type of force by reference to the table of enthalpies of vaporization ($kJ\ mol^{-1}$) given below.

Solid	He	Xe	HCl	Cl_2	CO_2	H_2O	Na	NaCl
$\Delta H_{vap}^{\ominus}$	0.10	15.0	21.1	26.8	26.0	50.0	107.8	222

$[\Delta H_{f(298\ K)}^{\ominus}(NaCl(s)) = -411\ kJ\ mol^{-1}$

$\Delta H^{\ominus}(\tfrac{1}{2}Cl_2(g) + e = Cl^-(g)) = -240\ kJ\ mol^{-1}$

First ionization energy of Na is $494\ kJ\ mol^{-1}$.

The heterolytic bond energy of Na—Cl(g) giving $Na^+(g)$ and $Cl^-(g)$ has been estimated to be $556\ kJ\ mol^{-1}$.]

(*Welsh S*)

12 'The particles in a substance (which may be atoms, molecules or ions) may be held together by ionic, covalent or hydrogen bonds, or by van der Waals' forces.'

(a) Illustrate this statement by reference to carbon, carbon dioxide, water, sodium chloride and a noble gas. Electron configurations should be given wherever possible.

(b) Relate the physical properties of the substances named in (a) to the type of bonding present.

(c) Suggest a reason why 2-nitrophenol, [structure of 2-nitrophenol with NO_2 and OH groups] is more volatile than its isomer, 4-nitrophenol, [structure of 4-nitrophenol with NO_2 and OH groups]

(*London*)

13 Discuss the types of bonding which occur in the following solid compounds:

(a) $CuSO_4 \cdot 5H_2O$, (b) C_6H_5COOH (do not discuss the bonding within the benzene ring), (c) P_4O_{10}.

(*Oxford and Cambridge S part question*)

14 (a) Describe, with illustrations, the types of crystal lattice present in graphite, diamond and sodium chloride.

By reference to the type of bonding present, explain the relative hardness and electrical conductivity shown by these substances.

(b) Sodium chloride exhibits *6,6-coordination* whilst caesium chloride shows *8,8-coordination*. Explain what these italicized terms mean.

(*AEB 1978 part question*)

15 Describe the use of X-rays to determine the structure of solids. Illustrate and discuss the different types of X-ray diffraction pattern that can be obtained and mention any limitations of the method.

X-ray studies of lithium chloride using X-rays of wavelength 0.0585 nm produced a strong diffraction at an angle of diffraction, θ, of 6.3° and another diffraction at 8.8°. There were also two related weaker diffractions at 5.4° and 10.9°. Use the Bragg diffraction equation to calculate the separation of the crystal planes in lithium chloride.

Draw suitable diagrams to show how the three sets of crystal planes are related to the unit cell of lithium chloride (face-centred cubic).

(*Nuffield S*)

16 'It is probably no exaggeration to say that in living processes the hydrogen bond is as important as the carbon—carbon bond.' Justify this statement by making detailed reference to a suitable range of examples.

(*Nuffield S*)

17 Give an account of the various bonds formed between ions, atoms, and molecules in the solid state.

(*Cambridge Entrance part question*)

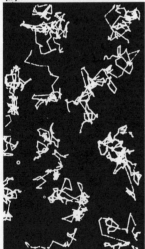

6

Matter in transition: changes of state

In this chapter we look at the structure of the most puzzling state of matter, the liquid phase. We see how a liquid forms when a solid melts, and go on to see what happens when the liquid evaporates and forms a gas. We shall find that a convenient way of discussing the conditions for the existence of different phases is in terms of phase diagrams, and we shall see what information some of them contain.

Introduction

Matter often comes in pools instead of lumps. Under the right conditions solids melt to liquids, and gases condense. In this chapter we look at ways of describing these changes of state, which are called *phase transitions*.

The liquid phase is the least understood of the phases of matter. This is because it is intermediate between a phase with a regular structure, such as a crystal, and a phase that is almost completely without structure, a gas. Nevertheless, progress has been made, and we shall look briefly at the modern description of liquids. Most of the discussion in this chapter, however, concerns how the phase transitions depend on the conditions. For instance, it is well known that the boiling temperature of water is lower at high altitudes than at sea level, and we shall show how to depict this dependence by drawing a *phase diagram*. Melting temperatures also depend on pressure (which is one reason why skaters move so easily over ice), and we shall explore the reasons.

6.1 The structures of liquids

We can arrive at a picture of the nature of liquids by thinking about what happens when solids melt and liquids boil.

Solids melting

When a solid is heated there is at first very little noticeable change. A block of copper at 100 °C looks much the same as one at 25 °C. Measurements, however, show that the block has expanded slightly. The reason for the expansion is easy to find: the ions are vibrating more vigorously and so are taking up more space. Computer simulations of the motion,

6.1 A computer simulation of a solid (a) just below and (b) just above its melting temperature

Fig. 6.1(*a*), show that although the vibrations swing the ions backwards and forwards they do not move away from their original lattice points.

As the temperature is raised the vibrations become more vigorous. The shape of the solid remains much the same because the motion swings the ions around their original locations, but the dimensions of the sample are greater. Finally, there comes a point at which the vibrations are so vigorous that the ions break away from their lattice points and the solid melts into a mobile liquid. The computer simulation of the motions of the ions just above the melting temperature, Fig. 6.1(*b*), shows that they are now free to move over large distances. Because its structure is so sloppy, a liquid adopts the shape of its container and flows easily. Liquids are much less compressible than gases because their particles are already close together and, unlike gases, there is little unfilled space.

Experimental evidence that molecules in liquids are in a state of constant motion was first put forward by Brown (in 1827). He observed the motion of small grains suspended in water and noticed through a microscope that they constantly jittered around. This *Brownian motion* fits in very well with the modern ideas on structure, because every grain is subjected to the battering of large numbers of molecules, but the battering is not uniform. For a short time the many impacts on one side may exceed those on the other, so that the grain is driven in that direction. Then another side receives most impacts, and so the grain is driven in another direction, Fig. 6.2.

Modern techniques (particularly special developments of spectroscopy) have displaced this type of crude observation, and we now have detailed information about molecular motion in liquids. These observations show that molecules in liquids move in a manner intermediate between the completely free motion of molecules in perfect gases and the almost total lack of motion in rigid crystals. When molecules move in liquids they do so in short steps, each step being through a fraction of the radius of the molecule (through about 10 pm). The steps are in random directions and occur every 10^{-13} s or so. Figure 6.1(*b*) should be interpreted with that in mind.

Liquids evaporating

Some molecules of a liquid can escape into the gas phase, and their number increases with the temperature. The pressure they exert is called the *vapour pressure* of the liquid at the temperature of the sample. The vapour pressure of water at 25 °C is only 3.2 kPa (0.03 atm) because only a small proportion of molecules can escape. When the temperature is 100 °C the vapour pressure rises to 101 kPa (1 atm). Under normal conditions this pressure drives back the atmosphere and the water evaporates. That is what happens when a kettle boils. The *normal boiling temperature* is the temperature at which the vapour pressure of a liquid reaches 101 kPa. Under different conditions (*e.g.*, up mountains and down mines) the vapour pressure matches the local atmospheric pressure at different temperatures, and so the boiling temperatures are different. Liquids more volatile than water reach vapour pressures of 101 kPa at

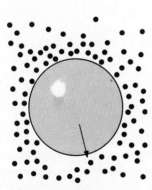

6.2 The molecular explanation of Brownian motion

lower temperatures. Liquid helium, for example, reaches that value when the temperature rises to only 4.2 K.

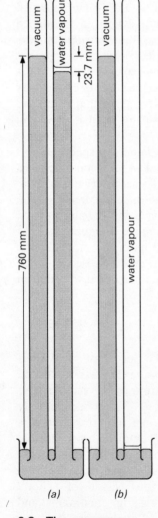

6.3 The vapour pressure of water at two different temperatures

Example: The vapour pressure of mercury at 25 °C is 0.224 Pa. What is the pressure in the 'vacuum' at the top of a mercury barometer tube? A few drops of mercury are spilt on the floor of a 30 m³ room. What mass of mercury will be found in the atmosphere within the room?

Method: Mercury evaporates until it exerts a partial pressure of 0.224 Pa. For the second part of the problem assume that the vapour acts as a perfect gas, and calculate the amount of substance in the atmosphere of the room using the perfect gas law in the form $n = pV/RT$. Convert to a mass on the basis that $A_r(\text{Hg}) = 200.6$.

Answer: The pressure in the 'vacuum' is the vapour pressure of the mercury, or 0.224 Pa. Inside a room of volume 30 m³ a partial pressure of mercury of 0.224 Pa corresponds to an amount of substance of Hg given by

$$n = \frac{pV}{RT} = \frac{(0.224 \text{ N m}^{-2}) \times (30 \text{ m}^3)}{(8.314 \text{ J K}^{-1} \text{ mol}^{-1}) \times (298 \text{ K})} = 2.71 \times 10^{-3} \text{ mol}.$$

(We have used $1 \text{ N} = 1 \text{ J m}^{-1}$.) This amount of substance corresponds to a mass

$$M = n \times (A_r \text{ g mol}^{-1}) = (2.71 \times 10^{-3} \text{ mol}) \times (200.6 \text{ g mol}^{-1}) = 0.54 \text{ g}.$$

Comment: 0.54 g of mercury vapour is dangerous: always clear up any spilt mercury. The vapour pressure of mercury is low, but not nearly low enough for experiments involving a very high vacuum. Other liquids have even lower vapour pressures. Tricresyl phosphate, for instance, has a vapour pressure of only 4.1×10^{-5} Pa at 25 °C. That corresponds to a column of mercury only one atom high! In modern high vacuum work liquids are not used at all.

6.2 Plotting the changes: phase diagrams

A simple way of measuring the vapour pressure of a liquid is to float some on to the surface of the mercury in a barometer tube, Fig. 6.3(a). The liquid evaporates into the vacuum until it reaches equilibrium with its own vapour. The resulting vapour pressure depresses the column of mercury, and the change of level gives its magnitude. At 25 °C, for example, the change of level caused by water is only 23.7 mm. At 100 °C the change is 760 mm, and the mercury inside the tube is level with the mercury outside, Fig. 6.3(b). The vapour pressures at a series of temperatures can be measured and the points plotted on a graph, Fig. 6.4. This is the start of the construction of a *phase diagram*.

The phase diagram of water
The complete phase diagram for water near room temperature and

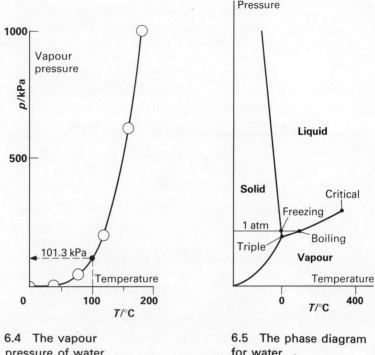

6.4 The vapour
pressure of water

6.5 The phase diagram
for water

pressure is shown in Fig. 6.5. The lines on it mark the conditions under
which different pairs of phases are in equilibrium. The line separating
liquid and vapour is the *vapour pressure curve* (it is Fig. 6.4 drawn on a
different scale). The line separating the solid and vapour regions is the
vapour pressure curve of ice. It marks the conditions where ice and water
vapour exist in equilibrium, and is called the *sublimation curve*. (A lump
of ice floating on mercury in a barometer tube evaporates until it is in
equilibrium with its vapour; at −5 °C its vapour pressure (400 Pa) is so
low that it will depress the level by only 3 mm.)

The almost vertical line in the diagram marks the conditions where ice
and liquid water are in equilibrium. It is the *melting temperature curve*
of ice. The slope of the line shows that the melting temperature falls as
the pressure is increased. When the pressure is 7×10^6 Pa (approximately the pressure exerted by a 70 kg ice skater with 1 cm^2 contact area) the
melting temperature is about −1 °C. This is because the high pressure
tends to crush the hydrogen bonds holding the water molecules in an
open structure, and so the denser liquid phase is formed.

There is one set of conditions where ice, liquid, and water vapour can
exist together in equilibrium. This is the *triple point*, Fig. 6.5. It occurs at
611 Pa (0.006 atm) and at exactly 273.16 K. There is no way of altering
these conditions while having all three phases present in equilibrium.
Because the point is unique it is used as a second fixed point in the
definition of the Kelvin temperature scale (Section 4.1). Another interesting point on the phase diagram is the *critical point*, where the vapour

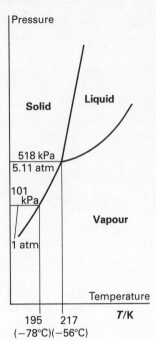

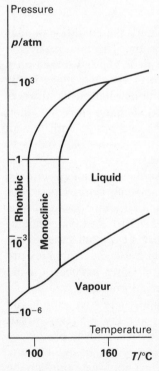

Pressure

Solid   Liquid

518 kPa
5.11 atm

101
  kPa

                 Vapour

1 atm

                          Temperature

195   217   T/K
(−78°C)(−56°C)

6.6 The phase diagram for carbon dioxide

Pressure

p/atm

—10³

—1

Rhombic   Monoclinic   Liquid

—10⁻³

                 Vapour

—10⁻⁶

                          Temperature

100   160   T/°C

6.7 The phase diagram for sulphur

pressure curve stops. Above the corresponding *critical temperature* the vapour and liquid are indistinguishable.

The phase properties of water make an appearance in everyday affairs in several ways. For instance, water vapour in the atmosphere can lead to the formation of frost in two ways. In one the first step is the formation of *dew*, when the temperature of damp air falls and the vapour condenses as a liquid. As the temperature falls still further the dew freezes, leading to a layer of *frost*. On very dry days, days when the vapour pressure of water is less than 611 Pa (the pressure of the triple point) lowering the temperature leads directly from the vapour to the solid region of the phase diagram. This gives rise to *hoar frost*, the direct formation of ice without an intermediate liquid state appearing.

The phase diagram of carbon dioxide

The experimentally determined phase diagram for carbon dioxide is shown in Fig. 6.6. There are several features that make it different from that of water. In the first place it shows that solid carbon dioxide sublimes if left in the open air (hence the name 'dry ice'). The liquid phase does not exist below 518 kPa (5.11 atm), and so at least that pressure must be exerted if the liquid is required. The vapour pressure of the solid is greater than 1 atm at all temperatures above −78 °C, and so it simply vaporizes if open to the atmosphere.

The other point to note is that increasing the pressure *increases* the melting temperature of carbon dioxide. This is because the solid is denser than the liquid since the small rod-like molecules pack together more closely. The effect of pressure is to favour the formation of the solid, and so it remains until higher temperatures. This behaviour is typical of most materials; water is unusual (yet again).

The phase diagram of sulphur

Sulphur can exist in different solid forms; these are called *allotropes* and the property is *allotropy* (the corresponding term for compounds is *polymorphism*). The phase diagram in Fig. 6.7 shows what conditions are needed in order to ensure that any given phase is stable and the conditions under which pairs of phases are in equilibrium. Note that sulphur has three different triple points. Other elements showing allotropy include tin, phosphorus, and carbon. The term *monotropy* is used to signify that only one allotrope is the stable form at all temperatures and 1 atm pressure (*e.g.*, phosphorus). *Enantiotropy* signifies that different forms are stable at various temperatures and 1 atm pressure (*e.g.*, sulphur). If the early polar explorers had understood phase diagrams they would not have taken their food in tin cans. The enantiotropy of tin gives rise to *tin plague*: the transition occurs at 13 °C but occurs rapidly only at temperatures well below 0 °C. As a result, the tin plate became porous, and the cans corroded. Nor would Napoleon's troops have lost their buttons at Moscow.

6.3 Other phases

There are three types of phase that do not readily fit into the scheme of

6.8 The partial order in a liquid crystal

solid, liquid, and gas. One unusual phase is a *plasma*. This is an ionized gas formed when high temperatures strip electrons from atoms. The importance of plasmas lies in their role throughout high-temperature physics, such as in stellar atmospheres, nuclear fusion research, and in rocket exhausts. Another unusual phase is a *glass*. Glasses are *supercooled liquids*; that is, liquids that have been cooled below their freezing temperatures but have not crystallized. Although apparently solid, structural studies show that glasses resemble liquids because their component particles are in a random jumble; they are *amorphous*, or formless. Glass does in fact flow, but only very slowly.

The third unusual phase is a *liquid crystal*. Although liquid crystals flow like ordinary liquids their molecules congregate in swarms showing some structure, Fig. 6.8. They are therefore intermediate in structure between true crystals and ordinary liquids. These peculiar properties are put to use in calculator and watch displays: the display depends on the orientating effect of very small electric potential differences and the resulting change in the optical properties of the liquid crystal. Their great advantage is that they consume very little power. One point about the second industrial revolution, the microelectronic revolution now in progress, is that, unlike the first, it is making extremely small demands on power and leading to the control rather than the generation of pollution.

Summary

1 *Phase transitions* are the changes of state of materials, and include melting and freezing, condensation and vaporization (evaporation), and changes of one solid phase into another.

2 The *temperature* of a phase transition depends on the *pressure*.

3 In solids, particles vibrate at their lattice points; at the melting temperature they are free to wander away.

4 In liquids molecules move in rapid small steps: this appears as the *Brownian motion* of suspensions of larger particles.

5 The *vapour pressure* of a liquid (or a solid) at a specified temperature is the pressure at which the liquid (or solid) is in equilibrium with its vapour.

6 The *normal boiling temperature* of a liquid is the temperature at which its vapour pressure is 1 atm.

7 A *phase diagram* shows the regions of temperature and pressure under which a phase is stable, and the lines show the conditions of temperature and pressure at which two phases exist in equilibrium.

8 The *triple point* is the condition of pressure and temperature at which three phases simultaneously exist in equilibrium.

9 *Sublimation* is the vaporization of a solid (and the reverse condensation) without an intervening liquid phase.

10 When the pressure is increased the melting temperature of ice is lowered. This is related to the contraction that occurs when ice melts.

11 When the pressure is increased on solid carbon dioxide its melting temperature rises. This is related to the expansion that occurs when the

solid melts. Most substances behave like carbon dioxide in this respect.

12 Different solid phases of elements are called *allotropes*. *Monotropy* signifies that only one allotrope is stable under 1 atm pressure at any temperature; *enantiotropy* signifies that different forms can be stable at various temperatures.

13 A *plasma* is an ionized gas.

14 A *glass* is a *supercooled liquid*. It is *amorphous* (formless) and not crystalline.

15 *Liquid crystals* flow like ordinary liquids but have molecules arranged with a degree of order.

Problems

1 Explain the terms *phase transition, vapour pressure, phase diagram, normal boiling temperature, melting temperature*.

2 Explain the terms *critical temperature* and *triple point*.

3 Why do the melting lines in the water and carbon dioxide phase diagrams slope in opposite directions? Which is more common?

4 Explain the terms *allotropy, monotropy*, and *enantiotropy*. Name four elements that show allotropy.

5 Account for the following properties of water: (*a*) ponds freeze from the surface down, (*b*) ice-skaters glide easily over ice, (*c*) a weighted wire passes through a block of ice (this is the phenomenon of *regelation*).

6 Account for the formation of (*a*) dew, (*b*) hoar frost, and (*c*) frost, and for (*d*) the non-existence of liquid carbon dioxide under atmospheric conditions.

7 The vapour pressure of water at 18 °C is 2.06 kPa. What depression of the column of mercury in a barometer would be observed if water is introduced and reaches equilibrium with its vapour?

8 What is the mass of water in the atmosphere of a 30 m^3 room when the relative humidity is 50% and the temperature 25 °C (when the vapour pressure is 3.17 kPa)?

9 The vapour pressure of carvone, which is responsible for the odour of spearmint, is about 15 Pa at 25 °C. How many molecules would be present in the atmosphere of a 30 m^3 room at equilibrium with a piece of chewing gum?

10 (*a*) Sketch the pressure–temperature phase diagram for water. Label your diagram to show the significance of the lines and areas in it.

    (*b*) Comment on the following statements.

        (*i*) It is very unusual for the melting-[temperature] of a pure substance to fall as pressure is applied.

        (*ii*) Water boils at a much higher temperature than hydrogen sulphide.

(Oxford part question)

11 (*a*) The following data refer to ammonia:

    Critical temperature 406 K;

    Critical pressure 1.13×10^4 kPa;

    Critical molar volume 0.72 dm^3 mol^{-1}.

Explain the meaning of each of these terms.

    (*b*) Methane cannot be liquefied at room temperature however great the pressure. Explain this statement.

(AEB 1980 part question)

7

Matter in mixtures

In this chapter we deal with the properties of simple non-reacting mixtures. We see how the boiling temperatures, freezing temperatures, and osmotic pressures of solutions depend on the amount of solute present, and how they can be used to determine molar masses. The behaviour of mixtures can be summarized neatly in terms of phase diagrams, and we see how to construct and interpret them. These diagrams can be used to discuss the techniques that are widely used for separating mixtures, both in the laboratory and in industry.

Introduction

Most of the materials we meet in the world are mixtures. In this chapter we investigate their properties, and look at the ways in which they differ from pure materials. The easiest way of depicting their properties is in terms of phase diagrams constructed experimentally. As well as summarizing experiments we should also be able to *predict* what happens when one substance is dissolved in another. For instance, we should be able to predict what happens to the freezing and boiling temperatures of a solvent when solute is added. In the process of seeing how to predict these effects we shall discover that the measurement of the properties of a mixture provides a way of determining the molar mass of the solute. The method is very useful for investigating the large molecules typical of polymers and biological materials.

7.1 Phase diagrams

The diagrams we consider are plots of temperature against composition. We shall normally express the compositions in terms of *mole fractions*, x (Section 4.1). The mole fractions of A and B in a two-component mixture are defined as

$$x_A = \frac{\text{amount of substance of A}}{\text{total amount of substance}} = \frac{n_A}{n_A + n_B}, \qquad x_B = \frac{n_B}{n_A + n_B};$$

n_A and n_B are normally expressed as so many moles of A or of B; x_A is a dimensionless number between 0 and 1. In a mixture of two components, x_A and x_B are related by $x_A + x_B = 1$.

Phase diagrams are determined experimentally by noting the temperatures at which changes occur in the nature of the system. For instance, we might note the boiling temperature of a liquid of some composition. Then repeating the observation for samples of different composition lets us draw the *liquid/vapour equilibrium line* on the phase diagram. Alternatively, we might note the temperature at which a solid precipitates out of solutions of various compositions.

Liquid/solid phase diagrams

Figure 7.1 shows a typical phase diagram for a system consisting of two components. Examples of this type of behaviour are benzene/naphthalene, potassium chloride/silver chloride, and tin/lead. The interpretation of the diagram can be understood by thinking about how it is constructed. Consider the point marked a in Fig. 7.2. This marks the melting temperature of pure A; point b marks the melting temperature of pure B. Now consider c_1. Under these conditions of temperature and composition the sample is entirely liquid. If the sample is allowed to cool its state is represented by the points lying on the vertical line below c_1. Point c_2 marks the temperature below which the sample is no longer entirely

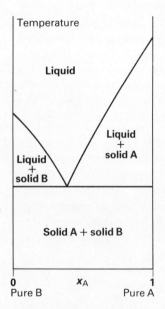

7.1 A typical solid/liquid two-component phase diagram

Principles of physical chemistry

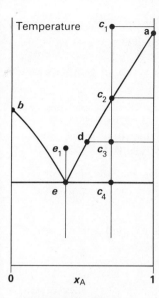

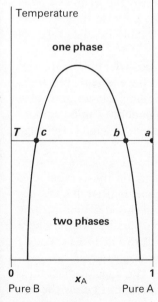

7.2 Changes represented by the phase diagram

7.3 The phase diagram for two partially miscible liquids

liquid; pure solid A begins to precipitate, leaving the liquid richer in B. Put another way, the point denotes the *solubility* of A in liquid B at the temperature corresponding to c_2.

When the temperature corresponds to c_3 the sample consists of some solid A plus liquid of composition d. On cooling the mixture still further, more solid A precipitates and the composition of the remaining solution becomes even richer in B. At the temperature corresponding to c_4 the last of the liquid (now of composition e) freezes. The solid then consists of regions of the two immiscible solids A and B, and these remain however low the temperature.

The composition corresponding to e is special. This can be appreciated by thinking about what happens when a sample is prepared with exactly that composition. At a high enough temperature, at e_1, the mixture is entirely liquid. As it is cooled it remains liquid, and no change is observed right the way down to e. At that temperature the entire solution solidifies into solid A plus solid B; at no stage does any A or B precipitate. *All* the solution solidifies at a *definite* temperature, just like a pure material. Furthermore, this temperature is lower than the freezing temperature of either pure component. A *mixture* (it is not a compound) showing this behaviour is called a *eutectic* (from the Greek for 'easily melted').

Eutectics have a number of commercial applications. For instance, tin and lead form a eutectic that melts at the conveniently low temperature of 183 °C; it is used as electrical solder. The eutectic mixture of common salt and water freezes at −21.2 °C, significantly below the freezing temperature of pure water; when salt is spread on wet roads, ice does not form until the temperature falls to well below 0 °C (but the spreading does not give the eutectic composition, except by chance).

Liquid/liquid phase diagrams

Figure 7.3 shows a typical phase diagram for a system composed of two *partially miscible liquids* (liquids that do not mix in all proportions at all temperatures). An example is nitrobenzene/propanone. The interpretation of the diagram follows from the way it is constructed experimentally.

The diagram refers to the behaviour of the system when it is at some constant pressure (such as 1 atm). Consider the line at the temperature T. Point a corresponds to pure A (*e.g.*, propanone). As substance B (*e.g.*, nitrobenzene) is added at constant temperature, the composition moves towards b. At b itself we notice a change. Instead of there being only a single liquid there are now two. The point b marks the composition at which the sample splits into two *phases*; one liquid phase is a saturated solution of B in A; the other is a saturated solution of A in B. In principle (by waiting long enough) the less dense phase floats on the other, but in practice the sample simply goes cloudy. The new layer is not pure B; B can dissolve A to some extent, and so the new phase is a saturated solution of A in B, with a composition denoted by c.

As still more B is added the new layer increases, but its composition, B saturated with A, remains constant. The *proportion* of the original layer decreases, but its composition, A saturated with B, remains constant.

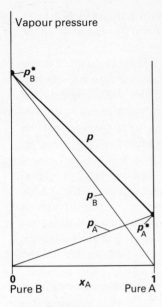

7.4 Raoult's law

When enough B has been added to bring the *overall* composition to c, it can dissolve all the A present. Therefore the two-phase liquid becomes a single phase again, and remains a single-phase liquid however much more B is added. Repeating the experiment at a series of temperatures then lets us draw the entire curve shown in Fig. 7.3.

Liquid/vapour phase diagrams

The French chemist Raoult spent much of his life measuring vapour pressures of liquid mixtures, and summarized his results (in 1886) as

Raoult's law: The partial vapour pressure of a component in a mixture is equal to the vapour pressure of the pure component multiplied by its mole fraction in the liquid.

If the vapour pressure of the pure component A is p_A^*, and A is present in a mixture with a mole fraction x_A, then it follows that

$$p_A = x_A p_A^*. \tag{7.1.1}$$

The total vapour pressure of a mixture of two liquids is the sum of their partial vapour pressures $p = p_A + p_B$. (This is a consequence of Dalton's law.) Therefore, if a mixture consists of A and B present with mole fractions x_A and x_B, the total vapour pressure is

$$p = x_A p_A^* + x_B p_B^* = x_A p_A^* + (1 - x_A)p_B^* = p_B^* + (p_A^* - p_B^*)x_A. \tag{7.1.2}$$

This shows, Fig. 7.4, that the vapour pressure of the mixture is p_B^* when A is absent ($x_A = 0$), and changes *linearly* as x_A increases. When B is absent ($x_A = 1$) the vapour pressure is p_A^*. Liquids that show this behaviour are called *ideal solutions.*

The linear dependence of the vapour pressure on the composition is found in practice for closely similar pairs of liquids because their van der Waals interactions are similar. This is illustrated in the case of benzene/methylbenzene in Fig. 7.5.

The *normal boiling temperature* of a liquid mixture is the temperature at which its total vapour pressure is 1 atm. Mixtures have vapour pressures that depend on their composition, and so the boiling temperature of a mixture also depends on its composition. The dependence can be worked out by considering how diagrams of the type shown in Fig. 7.4 change with temperature.

Figure 7.6 shows how the total vapour pressure of an ideal solution changes as the temperature is increased. The points on the vertical axes show the rising vapour pressures of pure A and pure B. Raoult's law has been used to draw the lines connecting them. The diagram can be converted into a boiling temperature/composition plot by noting where the lines cut the horizontal line at 1 atm. This leads to Fig. 7.6(*b*). All we have to do to find the boiling temperature of a mixture of given composition is to read off the corresponding temperature. This type of diagram is a *liquid/vapour phase diagram*, because at all temperatures below the line the stable state of the sample is as a liquid, and at all temperatures above the line the stable state is as a vapour. The importance of diagrams like this in real life is that they let us discuss distillation in a systematic way.

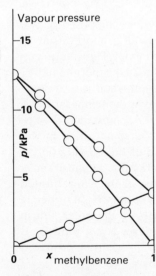

7.5 An almost ideal benzene/methylbenzene mixture

Principles of physical chemistry

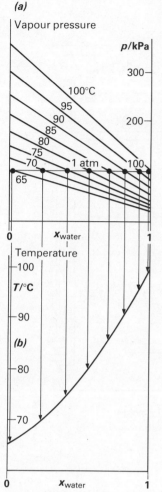

(a)

Vapour pressure

p/kPa

100°C
95
90
85
80
75
70
65
1 atm
100

300
200

0   x_{water} 1

Temperature
—100

T/°C

—90

(b)

—80

—70

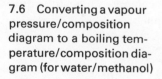

0   x_{water} 1

7.6 Converting a vapour pressure/composition diagram to a boiling temperature/composition diagram (for water/methanol)

Distillation

The vapour in equilibrium with a liquid mixture is richer in the more volatile component. We therefore need a second point on the phase diagram to denote its composition. Figure 7.7 shows how this can be done. Point a indicates the boiling temperature of a mixture of some composition. Point a' indicates the composition of the vapour in equilibrium with the liquid at that temperature. It is determined by withdrawing a little of the vapour when the sample is boiling steadily, and analysing it. The line joining the two points, one denoting the composition of the boiling liquid and the other the composition of its vapour, is called a *tie-line*. The entire vapour composition curve may be constructed by repeating the procedure with liquids of different compositions.

In distillation we expect the more volatile component to evaporate preferentially. In terms of the phase diagram, we start with the mixture with the initial composition $x_A(a)$ and heat it to boiling. The vapour has

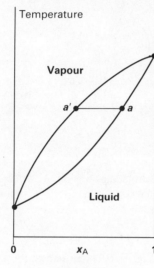

Temperature

Vapour

a' a

Liquid

0   x_A 1

7.7 A liquid/vapour temperature/composition phase diagram

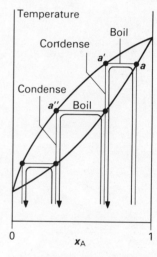

Temperature

Boil
Condense
a' a

Condense
a'' Boil

0   x_A 1

7.8 Fractional distillation

the composition $x_A(a')$ and is richer in B. If that vapour is removed and condensed, it gives a *condensate* that is richer in B than is the original mixture, Fig. 7.8. (If the less volatile component is a nonvolatile solid, only the solvent is distilled.)

In order to separate two volatile liquids entirely it is necessary to do a series of distillations. The condensate from the first distillation, of composition $x_A(a')$, is itself distilled. From the phase diagram, Fig. 7.8, we see that it boils to give a vapour of composition $x_A(a'')$, which is even richer in the more volatile component. This condensate may be distilled again, and so on. After several such steps a liquid rich in B is obtained. In practice this process of *fractional distillation* is carried out in a single piece of apparatus, which may be of laboratory, Fig. 7.9, or of industrial

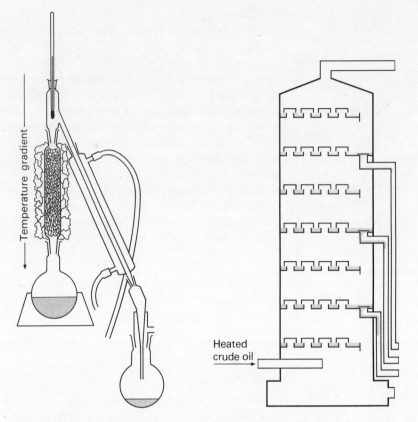

7.9 A laboratory-scale fractional distillation apparatus

7.10 Industrial-scale fractional distillation

Heated crude oil

Temperature gradient

dimensions, Fig. 7.10. The sequence of boiling and condensation steps takes place on the plates or the packing of the columns where the rising vapour meets the falling liquid and reaches equilibrium.

Deviations from ideality

Mixtures of some liquids have vapour pressures greater than Raoult's law predicts; these *non-ideal solutions* show *positive deviations*, Fig. 7.11. The molecules in the mixture have unfavourable interactions with each other, and a greater tendency to vaporize than in an ideal solution. A higher vapour pressure corresponds to a lower boiling temperature, and so mixtures showing positive deviations from Raoult's law have lower boiling temperatures than expected. Sometimes the deviations are so great that the boiling temperature curve passes through a minimum, and the mixture boils at a temperature lower than either pure component. Mixtures showing this behaviour are said to form a *minimum boiling azeotrope*. Ethanol and water form an azeotrope of this kind, Fig. 7.12; the minimum boiling temperature is 78.2 °C, and occurs when the composition corresponds to 95.6% ethanol.

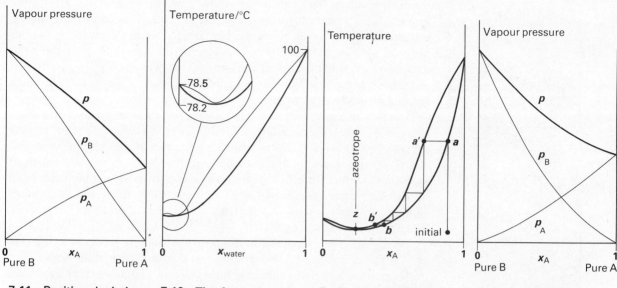

7.11 Positive deviations from Raoult's law

7.12 The formation of a minimum boiling azeotrope (ethanol/water)

7.13 Fractional distillation of an azeotropic mixture

7.14 Negative deviations from Raoult's law

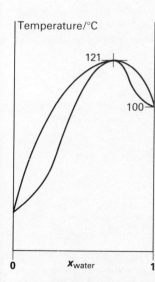

7.15 The formation of a maximum boiling azeotrope (nitric acid/water)

The reason for the name *azeotrope* (which comes from the Greek for 'boiling without change'), and its importance for distillation, can be understood by thinking about distilling a mixture at composition $x_A(a)$, Fig. 7.13. This mixture boils at the temperature corresponding to a, and the vapour, point a', is richer in ethanol. Condensation of the vapour and redistillation shifts its composition further towards ethanol. At b, however, the compositions of the vapour and liquid are almost exactly the same (b' is close to b). At z they are exactly the same. A liquid of composition $x_A(z)$ boils to give a vapour of exactly the same composition (hence 'boils without change'). Distillation cannot concentrate the mixture beyond this *azeotropic composition*. Anhydrous ethanol therefore has to be prepared by chemical drying, such as by shaking with silica gel.

Some pairs of liquids form mixtures having vapour pressures lower than Raoult's law predicts, Fig. 7.14. These non-ideal solutions show *negative deviations* on account of the favourable interactions between the components. Lower vapour pressure implies a higher boiling temperature, and so such mixtures boil at higher temperatures than expected. When the deviations are very large the boiling temperature may be higher than either pure component, and a *maximum boiling azeotrope* is obtained, Fig. 7.15. An example is 68.2% nitric acid, which boils unchanged at 121 °C.

Immiscible liquids

When the two liquids are completely immiscible, like oil and water, the total vapour pressure of an agitated sample is the sum of their individual

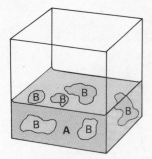

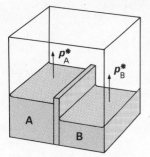

7.16 The vapour pressure of immiscible liquids

vapour pressures. This can be understood on the basis of the illustration in Fig. 7.16: a sample of two immiscible liquids is equivalent to two separated liquids inside the same container. Liquid A exerts its vapour pressure (p_A^*) and liquid B exerts its vapour pressure (p_B^*); the total vapour pressure is the sum of the two vapour pressures, $p_A^* + p_B^*$, and it is the same whether the liquids are separated or one is a collection of blobs floating around in the other.

This behaviour of immiscible liquids is the basis of *steam distillation*. Since a sample of immiscible liquids exerts the sum of their vapour pressures, it boils at a lower temperature than does either component alone. Therefore a sample can be distilled (by bubbling steam through it) at a lower temperature, and temperature-sensitive organic molecules can be distilled under conditions that do not cause them to decompose. Boiling cabbages results in the steam distillation of some of their more disagreeable constituents.

Steam distillation was used in the past to obtain molar masses; but it is very inaccurate, messy, and wholly unnecessary now that mass spectrometry, osmosis, and other highly developed techniques are available. The following *Example* summarizes the procedure.

Example: A sample of phenylamine was steam distilled at 98 °C. The condensate was analysed and found to consist of 42.1 g of water and 16.3 g of the amine. At 98 °C the vapour pressure of the amine is 7.065 kPa and that of water is 94.260 kPa. What is the molar mass of the amine?

Method: Denote the phenylamine as P and the water as W. The composition of the condensate is the composition of the vapour in equilibrium with the boiling sample. Express the mass composition as mole fractions, using the molar masses of the two components (that of P is unknown). The partial pressures in the vapour are related to the total pressure p by $p_P = x_P p$ and $p_W = x_W p$. But we know the partial pressures (the vapour pressures of the individual pure components) and we also know the total pressure (their sum). Therefore the only unknown quantity in the calculation is $M_r(P)$. Solve the resulting equations for it. Use $M_r = M_m/\text{g mol}^{-1}$.

Answer: Amount of substance of W: $n_w = (42.1 \text{ g})/(18.02 \text{ g mol}^{-1}) = 2.34$ mol

Amount of substance of P: $n_P = (16.3 \text{ g})/M_m = (16.3/M_r)$ mol

Total amount of substance: $n_w + n_P = 2.34 \text{ mol} + (16.3/M_r)$ mol

Mole fraction of water: $x_W = \dfrac{2.34}{2.34 + (16.3/M_r)}$

Total vapour pressure at 98 °C: $p = (7.065 + 94.260)$ kPa
$= 101.325$ kPa

Mole fraction of water: $x_W = p_W/p = \dfrac{94.260 \text{ kPa}}{101.325 \text{ kPa}} = 0.9303.$

On equating the two expressions for x_W we have

$$\frac{2.34}{2.34 + (16.3/M_r)} = 0.9303, \quad \text{or} \quad 2.34 = 2.18 + (15.2/M_r).$$

Principles of physical chemistry

Therefore

$$M_r = \frac{15.2}{(2.34 - 2.18)} = 95.$$

It follows that the molar mass of the amine is 95 g mol^{-1}.

Comment: The formula for phenylamine is $C_6H_5NH_2$, and so the calculated molar mass is 93.13 g mol^{-1}. The inaccuracy of the method lies in the last step of the calculation: the denominator $2.34 - 2.18 = 0.16$ is the small difference between two numbers, and its value strongly affects the outcome.

7.2 Colligative properties

Phase diagrams are constructed on the basis of experimental observations. When the diagrams are inspected it is found that there are several regularities. For instance, it is found that the presence of a solute always lowers the freezing temperature of a solvent (which is why we use anti-freeze) and, if it is non-volatile, always raises the solvent's boiling temperature (oceans evaporate more slowly than do ponds).

Raoult's law is particularly simple when one of the species, B, is not volatile (or, at least, much less volatile than is the other component at the temperature of interest; everything is volatile at high enough temperatures). Since B is not volatile, its vapour pressure is effectively zero at the temperature of interest, and so we may set $p_B^* = 0$ in eqn (7.1.2). Then the vapour pressure of a solution of a non-volatile solute in a volatile solvent (A) is

$$p = x_A p_A^* \quad \text{or} \quad \frac{p_A^* - p}{p_A^*} = 1 - x_A = x_B. \tag{7.2.1}$$

This formula shows that there is a decrease in vapour pressure, from p_A^* to p, proportional to the mole fraction of the non-volatile solute. This is illustrated in Fig. 7.17. Since a lower vapour pressure implies a higher boiling temperature, we can also conclude that *the presence of a non-volatile solute raises the boiling temperature of a solution.* Notice that the vapour pressure lowering depends on the mole fraction of the *solute*, x_B, but is independent of its chemical nature. Such properties are called *colligative.* Thus, if we add N molecules of naphthalene to benzene we get the same decrease in vapour pressure as when we add N molecules of anthracene.

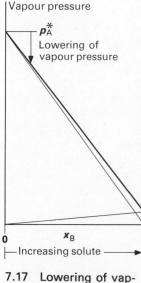

7.17 Lowering of vapour pressure by an (almost) non-volatile solute (B)

The elevation of boiling temperature

A pure solvent gives rise to a vapour pressure on acccount of the tendency of its molecules to escape and to disperse as a gas, Fig. 7.18(*a*). A gas is a chaotic collection of particles. When a non-volatile solute is added to a solvent the solution is more chaotic than the original solvent, Fig. 7.18(*b*). Since the solution is already chaotic as a result of the

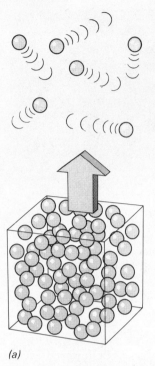

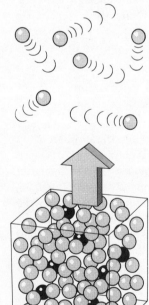

7.18 Vaporization of
(a) pure solvent,
(b) solution

presence of the solute, there is a lower tendency for the solvent molecules to escape as a gas. That is, the vapour pressure is lower. (The role of chaos in determining the direction of natural change, such as vaporization, can be made precise; it is developed in the next few chapters and explained in Chapter 11.)

The measurement of the elevation of boiling temperature is a way of determining the molar mass of the solute. The procedure is called *ebullioscopy*. When some simplifications are made, principally that the solution is dilute, it turns out that the elevation of boiling temperature, δT, is proportional to the molality (p. 238) of the solution, m, the relation being,

$$\textit{Elevation of boiling temperature}: \quad \delta T = K_b m \qquad (7.2.2)$$

K_b is called the *ebullioscopic constant*, and depends on the solvent; some values are listed in Table 7.1. The method of using this expression is described in the following *Example*.

Table 7.1 Cryoscopic and ebullioscopic constants

	$T_f/°C$	$\dfrac{K_f}{(K/mol\,kg^{-1})}$	$T_b/°C$	$\dfrac{K_b}{(K/mol\,kg^{-1})}$
Benzene	5.5	5.12	80.1	2.67
Camphor	179.5	40.0	208.3	6.0
Ethanoic acid	16.6	3.90	118.1	3.07
Water	0.0	1.86	100.0	0.52

Example: The normal boiling temperature of benzene is 80.1 °C and its ebullioscopic constant is 2.67 K/(mol kg^{-1}). When 0.668 g of camphor is dissolved in 20 g of benzene the boiling temperature of the solution is raised by 0.646 K (*differences* of boiling temperature can be measured more precisely than the boiling temperatures themselves). What is the molar mass of camphor?

Method: Equation (7.2.2) is used to find the molality of the solution: 0.668 g in 20 g of solvent corresponds to 33.4 g in 1 kg. A mass of 33.4 g corresponds to an amount of substance $(33.4\,g/M_m) = (33.4/M_r)$ mol, and the molality of the solution is therefore $(33.4/M_r)$ mol kg^{-1}. On equating the two expressions for the molality the only unknown is M_r; hence this may be found. Finally use $M_m = M_r\,g\,mol^{-1}$.

Answer: The molality of the solution is obtained from eqn (7.2.2) as

$$m = \delta T/K_b = \frac{(0.646\ K)}{[2.67\ K/(mol\ kg^{-1})]} = 0.242\ mol\ kg^{-1}.$$

The molality is also given as $(33.4/M_r)$ mol kg^{-1}. Equating these two expressions leads to

Principles of physical chemistry

$$M_r = \frac{33.4}{0.242} = 138.$$

Therefore, $M_m = 138 \text{ g mol}^{-1}$.

Comment: The formula for camphor is $C_{10}H_{16}O$; therefore its calculated molar mass is $152.23 \text{ g mol}^{-1}$, and so the result is inaccurate. This technique is rarely used for molar mass determinations because elevations of boiling temperature are very small and some materials are temperature-sensitive.

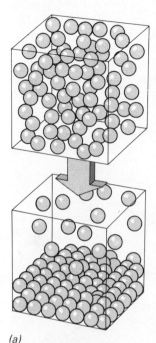

(a)

(b)

7.19 Freezing of (a) pure solvent, (b) solution

The depression of freezing temperature

We have seen that the presence of a solute makes a solvent more chaotic. As well as lowering the vapour pressure there is another consequence. Melting is similar to evaporation because when it occurs the particles of a solid collapse into a less regular arrangement. Freezing is the converse; the chaos of the liquid gives way to the orderliness of the solid. Figure 7.19 illustrates what happens when there is a solute present that does not dissolve in the solid. In the pure solvent an ordered structure comes about at a fixed temperature, the normal freezing temperature, Fig. 7.19(a). When solute is present there is even more randomness than in the original pure solvent, and so the temperature has to be lowered even further in order to bring about the transition to a more orderly form. In other words, *the presence of a solute lowers the freezing temperature of the solution*. This is illustrated in Fig. 7.19(b).

The depression of freezing temperature depends on the mole fraction of solute present and its measurement can be used to determine the latter's molar mass. The technique is called *cryoscopy*. After some calculation and simplification on the basis that the solution is dilute, it turns out that the depression of freezing temperature of a solution of molality m is

Depression of freezing temperature: $\delta T = K_f m.$ \hfill (7.2.3)

K_f is called the *cryoscopic constant*, and some values are included in Table 7.1. The value of K_f for camphor is large, and so it is frequently used in cryoscopy (when it is called *Rast's method*). Freezing-temperature depressions are generally greater than boiling-temperature elevations, and so cryoscopic values of molar masses are more reliable than are the ebullioscopic values. The *Example* that follows illustrates how eqn (7.2.3) is used.

Example: The freezing-temperature depression produced by dissolving 2.16 g of sulphur in 100 g of a solvent was 0.58 K. When 3.12 g of iodine was dissolved in 100 g of the same solvent the depression was 0.85 K. On the grounds that it is known that iodine is present as I_2 in this solvent, deduce the molecular form of the sulphur.

Method: The freezing-temperature depressions are given by eqn (7.2.3) in each case, with the same value of K_f. Therefore the ratio of the two depressions is equal to the ratio of the molalities of the two solutions. Since the mass of solvent is the same in each case, the ratio of the molalities is equal to the ratio of the amounts of substance in each case. The amounts of substance are given by (mass)/M_m in each case. Therefore

$$\frac{\delta T(\text{sulphur})}{\delta T(\text{iodine})} = \frac{n(\text{sulphur})}{n(\text{iodine})} = \left(\frac{\text{mass (sulphur)}}{M_m(\text{sulphur})}\right)\left(\frac{M_m(\text{iodine})}{\text{mass (iodine)}}\right).$$

Since the only unknown in this expression is the molar mass of the sulphur, it can be determined from the data. The relative atomic mass is known (32.06) and so the number of atoms per molecule can be determined. The relative atomic mass of I is 126.9 and so $M_m(\text{iodine}) = M_m(I_2) = 253.8$ g mol^{-1}.

Answer: Substitution of the data into the expression just derived gives

$$\frac{\delta T(\text{sulphur})}{\delta T(\text{iodine})} = \frac{0.58 \text{ K}}{0.85 \text{ K}} = 0.68.$$

$$\left(\frac{\text{mass(sulphur)}}{M_m(\text{sulphur})}\right)\left(\frac{M_m(\text{iodine})}{\text{mass(iodine)}}\right) = \left(\frac{2.16 \text{ g}}{M_m}\right)\left(\frac{253.8 \text{ g mol}^{-1}}{3.12 \text{ g}}\right)$$
$$= (176 \text{ g mol}^{-1})/M_m.$$

Therefore $M_m = (176 \text{ g mol}^{-1})/0.68 = 260$ g mol^{-1}. This molar mass corresponds to a molecular formula S_8; $M_r(S_8) = 256.5$.

Comment: Sulphur often exists as S_8, with the atoms forming a puckered ring resembling a crown.

Osmosis

There is another important colligative property that not only is very useful for the measurement of molar mass but also is of great biological importance. This is the property of *osmosis* (from the Greek for 'push'). Osmosis is the tendency of solvent molecules to pass through a membrane from a more dilute to a more concentrated solution. The membrane is special in the sense that it is *semi-permeable*; that is, it allows solvent but not solute molecules to pass through. In some cases the semi-permeable membrane may allow water to pass but not dissolved ions; in others it may allow a solvent composed of small molecules to pass but not large polymer or biological molecules. Cellulose ethanoate ('cellulose acetate') is often used as a membrane. The cytoplasm that lines cell walls is a naturally occurring semi-permeable membrane, which is why osmosis is important in biology.

When a dilute solution is brought into contact with a concentrated solution through the semi-permeable membrane, Fig. 7.20, the solvent flows through. As it does so, the height of the concentrated solution increases, and this gives rise to an opposing pressure. As the flow of solvent continues there comes a point when the extra pressure on the right of the membrane is great enough to balance the net flow. At that

h

Solvent | Solution

7.20 The development of osmotic pressure ($\Pi = \rho g h$)

Principles of physical chemistry

stage the two sides of the membrane are in equilibrium, and the net flow is zero. The excess pressure at equilibrium is called the *osmotic pressure*, and is denoted Π (Greek capital pi). The same process accounts for the uptake of water by the root systems of plants. An egg out of its shell swells when immersed in water, but shrinks if immersed in strong salt solution (the shell membrane is semi-permeable). 'Chemical gardens' are produced by dissolving various metal salts in water-glass (sodium silicate) solution. The metal silicates formed by double decomposition act as semi-permeable membranes – water passes through and into them, causing them to swell and distort into weird shapes. If a pressure *greater* than Π is exerted on a concentrated solution, the solvent passes back through the membrane to the dilute side. This *reverse osmosis* is used to purify sea water.

Osmosis is a colligative property because the osmotic pressure of a solution depends on the number of solute particles present and is independent of their nature. When a solution containing an amount of substance n of solute in a volume V is in contact with the pure solvent at a temperature T, the osmotic pressure is given by the *van't Hoff equation*

$$\Pi V = nRT, \tag{7.2.4}$$

where R is the molar gas constant. Note the striking similarity of the van't Hoff equation to the perfect gas equation ($pV = nRT$) even though it applies to an apparently quite different system.

Osmosis is very useful for the determination of molar mass. Measurements of osmotic pressure can be made at room temperature, and so fragile, temperature-sensitive biological molecules can be studied. It is also a very sensitive technique, and so can be applied to the determination of the molar masses of proteins, polymers, and other large molecules. Thus a $1\ \mathrm{mol\ dm^{-3}}$ solution of sucrose causes an elevation of boiling temperature of water of about 0.5 K, which is just measurable, a depression of freezing temperature of 1.9 K, which is significant but not dramatic, but exerts an osmotic pressure sufficient to sustain a column of water 250 m high!

Example: 1.10 g of a protein was dissolved in 100 g of water at 20 °C. The osmotic pressure of the solution was found to be 396 Pa. What is the molar mass of the protein?

Method: Use eqn (7.2.4) to find the amount of substance of the protein in the solution, then relate that to the molar mass through $n = \mathrm{mass}/M_m$. Note that $1\ \mathrm{Pa} = 1\ \mathrm{N\,m^{-2}} = 1(\mathrm{J\,m^{-1}})\mathrm{m^{-2}} = 1\ \mathrm{J\,m^{-3}}$. Insert the data $T = 293$ K and $\Pi = 396$ Pa into eqn (7.2.4). Since the density of the solution is almost exactly $1\ \mathrm{g\,cm^{-3}}$, take the volume as $V = 100\ \mathrm{cm^3} = 100 \times 10^{-6}\ \mathrm{m^3}$.

Answer:

$$n = \frac{\Pi V}{RT} = \frac{(396\ \mathrm{J\,m^{-3}}) \times (10^{-4}\ \mathrm{m^3})}{(8.314\ \mathrm{J\,K^{-1}\,mol^{-1}}) \times (293\ \mathrm{K})} = 1.63 \times 10^{-5}\ \mathrm{mol}.$$

The mass of the sample is 1.10 g; therefore the molar mass of the protein is

$$M_m = \frac{mass}{n} = \frac{1.10 \text{ g}}{(1.63 \times 10^{-5} \text{ mol})} = 6.7 \times 10^4 \text{ g mol}^{-1}.$$

Therefore the relative molecular mass is 67 000.

Comment: Osmosis is a very important technique for determining the molar masses of biological and synthetic macromolecules. It is widely used in modern laboratories. Note that an osmotic pressure of 396 Pa is sufficient to support a column of water 4.04 cm high, and so it is readily measurable in an apparatus like that shown in Fig. 7.20. In the *Hartley-Berkeley osmometer* an excess pressure is exerted on the concentrated solution so as to oppose the net flow of the solvent into it. The pressure required to stop the net flow is equal to Π. This method has the advantage of leaving concentrations unchanged, and the results are more accurate.

Summary

1 A *phase diagram* is a plot of pressures and compositions, or temperatures and compositions, at which phases coexist in equilibrium.

2 *Partially soluble materials* consist of solutes that are not soluble in the solvent in all proportions at all temperatures in the range of interest.

3 A *liquid/solid equilibrium line* denotes the *solubility* of one component in the other. The composition at a point on the line is the composition of the *saturated solution* at that temperature.

4 The *eutectic* composition is the composition corresponding to the lowest melting temperature of a mixture of solids, or the lowest freezing temperature of a liquid mixture. A mixture with the eutectic composition melts and freezes at a single temperature (at a stated pressure) without change of composition.

5 *Partially miscible liquids* are pairs of liquids that do not mix (to give a single liquid phase) in all proportions at all temperatures in the range of interest.

6 *Raoult's law* states that $p_A = x_A p_A^*$, eqn (7.1.1).

7 *Ideal solutions* are solutions that obey Raoult's law.

8 Solutions having vapour pressures lower than predicted by Raoult's law are said to show *negative deviations*; when the deviations are large enough the system forms a maximum boiling azeotrope. Higher vapour pressures than are predicted by Raoult's law are called *positive deviations* and, when large enough, lead to minimum boiling azeotropes.

9 When a liquid mixture is boiling, the equilibrium compositions of the liquid and the vapour are given by the two points at the opposite ends of the *tie-line* on the liquid/vapour phase diagram.

10 *Fractional distillation* is a series of distillations and condensations; it is used to separate mixtures of liquids of similar boiling temperature.

11 An *azeotrope* is a mixture that boils without change of composition. A *maximum boiling azeotrope* boils at a higher temperature than either

component; a *minimum boiling azeotrope* boils at a lower temperature than either component.

12   Azeotropic liquid mixtures cannot be completely separated by distillation.

13   Pairs of *immiscible liquids* boil at lower temperatures than either component alone, the total vapour pressure being the sum of the individual vapour pressures.

14   *Steam distillation* is used to separate temperature-sensitive materials, and was used for the determination of molar masses.

15   *Colligative properties* are properties that depend on the number but not on the nature of solute particles.

16   Colligative properties include the *elevation of boiling temperature*, the *depression of freezing temperature*, and *osmosis*.

17   The presence of *non-volatile solutes* lowers the vapour pressure of a solution, and therefore *raises the boiling temperature*; this is the basis of *ebullioscopy*, which is used for the determination of molar mass.

18   The presence of a solute that does not dissolve in the solid solvent leads to a *lowering of the freezing temperature* of the solution. This is the basis of *cryoscopy*, which is also used for the determination of molar mass.

19   *Osmosis* is the tendency for solvent to flow from a less to a more concentrated solution when the two are separated by a *semi-permeable membrane*.

20   The *osmotic pressure* is the extra pressure that must be exerted on the concentrated solution in order to stop the net flow by osmosis; its magnitude is given by the *van't Hoff equation*, eqn (7.2.4).

Problems

1   Explain the term *mole fraction*. Calculate the mole fractions of ethanol and water in vodka, taken to consist of 35% by mass of ethanol and 65% by mass of water.

2   Explain the terms *eutectic, azeotrope, minimum boiling azeotrope*, and *tie-line*.

3   Show how a temperature–composition phase diagram may be used to describe the fractional distillation of a mixture of two liquids.

4   The separation of methanol and ethanol is an important problem, not only on account of their different physiological effects. Their vapour pressures at 25°C are 11.730 kPa and 5.866 kPa, respectively. On the basis that 40 g of methanol and 200 g of ethanol mix to give an ideal solution, calculate (*a*) the partial vapour pressure exerted by each component, (*b*) the total vapour pressure of the mixture, (*c*) the composition of the vapour.

5   At 95 °C the vapour pressure of water is 84.52 kPa and that of bromobenzene (C_6H_5Br) is 15.48 kPa. Calculate the mass of water needed to steam distil 100 g of bromobenzene when the atmospheric pressure is 100.0 kPa.

6   What is the effect on the vapour pressure of water and the total vapour pressure of the system when small amounts of the following materials are added (separately) to water at 25 °C: (*a*) sodium chloride, (*b*) propanone, (*c*) tetrachloromethane?

7   Explain why the presence of a non-volatile solute raises the boiling temperature of a solvent. Explain why the presence of a solute that is insoluble in the solid solvent lowers the freezing temperature of the solution.

8   What is the molar mass of a substance which, when 0.2 g is dissolved in 10 g of water, lowers the freezing temperature to −0.62 °C?

9   Calculate the boiling temperature under a pressure of 1 atm of 100 g of water sweetened with 1 lump (3.6 g) of sucrose. What is the freezing temperature of the solution?

10   Calculate the osmotic pressure of the sucrose solution in the last question, at 25 °C (take the density of water as 1.00 g cm^{-3}).

11   Suggest an explanation for the similarity between the van't Hoff equation $\Pi V = nRT$ and the perfect gas equation $pV = nRT$.

12   What do you understand by the terms *allotropy*, *alloy*, and *eutectic mixture*? How could you distinguish between a eutectic mixture and a pure substance?

A mixture of 27% gold and 73% thallium melts and freezes at 131 °C as if it were a pure substance. Draw on graph paper the phase diagram for the gold–thallium system, labelling each of the main areas.

Sketch the temperature–time curves that would be obtained on allowing the following mixtures to cool to room temperature from 1000 °C:

	(a)	(b)
gold	60%	27%
thallium	40%	73%

Describe what happens in the system during the cooling of mixture (a).

(Oxford and Cambridge)

13   State Raoult's law as applied to a liquid binary mixture and describe how it can be explained in terms of a simple molecular model.

Show how binary liquid systems which deviate from ideality may form either a maximum or a minimum boiling mixture.

Discuss the principles involved in the separation, by fractional distillation, of a mixture of two volatile liquids (which do not form a constant boiling mixture). Give brief details of the apparatus and methods you would use to carry out this separation in the laboratory.

(Oxford and Cambridge)

14   (a) Draw and suitably label a phase diagram for water. Use the diagram to show the effect of pressure on the freezing point [temperature] of pure water.

Explain how the vapour pressure changes when an involatile solute is added to water. Add a further part to your diagram to illustrate the effect of this addition on the boiling point and freezing point of pure water.

(b) When an immiscible volatile liquid is added to water explain how the total vapour pressure is affected. Use this concept to explain the basis of steam distillation and give one practical application of steam distillation.

(c) Hexane and heptane are totally miscible and form an ideal two component system. If the vapour pressures of the pure liquids are 56 000 and $24\,000 \text{ N m}^{-2}$ at 51 °C calculate (i) the total vapour pressure and (ii) the mole fraction of heptane in the vapour above an equimolar mixture of hexane and heptane.

(Welsh)

15   (a) Explain the meaning of the term *osmotic pressure*. Describe with the aid of a sketch an apparatus for the measurement of the osmotic pressure of an aqueous solution.

(b) At 290 K, the osmotic pressure of a solution of 24.3 g of a sugar in 1 dm³

Principles of physical chemistry

of water is 9.85×10^4 Pa. Estimate the relative molecular mass (molecular weight) of this sugar. ($1\,Pa = 1\,N\,m^{-2}$.)

(*c*) When the relative molecular mass of a compound is comparatively low, it is far more often measured by determination of freezing-point [temperature] depression or boiling-point elevation than by osmotic pressure measurements. For macromolecules such as polymers, the reverse is true. Why is this?

(Oxford)

16 Describe how measurements of freezing point [temperature] depression can be used to determine relative molecular masses (molecular weights). It was found that 1 mole of added solute depressed the freezing point of 1 kg of benzene by 5.10 K, while 0.500 g of a hydrocarbon M dissolved in 100 g of benzene depressed the freezing point by 0.212 K. Quantitative analysis indicated that the composition was 90.0% carbon and 10.0% hydrogen by mass. A spectroscopic method established that M possessed a threefold axis of symmetry. Suggest a possible structure for M. What physical method might be used to provide evidence about the symmetry of a molecule?

(Oxford Entrance)

17 (*a*) Sketch the boiling [temperature]/composition diagram for a simple binary mixture which *does not* show a boiling [temperature] maximum, and use it to illustrate the principle of fractional distillation.

(*b*) Explain *concisely* how steam distillation works.

(*c*) Methanol and ethanol form a solution which is almost ideal, and the vapour pressures of the pure components at 300 K are $11.82 \times 10^3\,N\,m^{-2}$ and $5.93 \times 10^3\,N\,m^{-2}$, respectively. Calculate the mole fraction of methanol in the vapour above a solution made by mixing equal *weights* of the two components.

(*d*) The vapour pressure of the *immiscible* liquid system diethylaniline–water is $10.13 \times 10^4\,N\,m^{-2}$ at 372.5 K. The vapour pressure of water at this temperature is $9.92 \times 10^4\,N\,m^{-2}$. How many kg of steam are required to distil over 0.1 kg of diethylaniline at a total pressure of $10.13 \times 10^4\,N\,m^{-2}$?

(Cambridge Entrance)

8

Ions in solution

In this chapter we examine the properties of a special class of solution, one containing ions. The principal difference between these and the solutions considered in the last chapter is that the solute particles carry charges, and so the solutions conduct electricity. We shall see how the presence of ions affects the properties of solutions, and pay special attention to the description of their electrical properties.

Introduction

A special type of solution is formed by dissolving an *electrolyte*, a substance able to act as a supply of *ions*. While solutions of electrolytes show all the colligative properties discussed in the last chapter, they have the additional and very important property of being able to conduct electricity.

8.1 Solutions of electrolytes

When the view was expressed by Arrhenius (in 1884) that many materials simply fell apart into ions when they dissolved, his contemporaries were very sceptical, Box 8.1. After all, they argued, chemical bonds are strong, and the gentle act of dissolution could not be expected to break them. We now know that there are two errors in this argument.

Box 8.1: The development of the concept of ions

Baron von Grotthuss took the view that the 'molecules' of electrolytes were polar, and that the application of an electric field caused them to dissociate and to give up their electric charge to the electrodes. There was a shift of opinion with Rudolph Clausius (b. 1822 in the part of Prussia that is now Poland) who thought that electrolytes were partly dissociated even in the absence of an electric field. Svante Arrhenius (born in Sweden in 1859) put forward the suggestion that in solution electrolytes were completely dissociated into *ions* (a term that had been coined by Michael

Faraday). He did so in his doctoral thesis; it was greeted with so much disbelief that he was awarded a mere fourth class pass. Nevertheless, evidence accumulated in its favour with the work of J. H. van't Hoff (b. 1852 in Rotterdam, and a student of Kekulé), who later received the first Nobel prize in chemistry (in 1901) for his work on equilibria, kinetics, and osmosis. The theory of complete dissociation was taken up and supported by the Latvian chemist Wilhelm Ostwald, who can be said to be the father of physical chemistry because he did so much to organize it into a coherent subject. The terms ion, cation, anion, cathode, anode, and electrode were all coined by Faraday who began life (in Surrey and in 1791) as the son of a blacksmith, became the assistant of Humphry Davy, and in due course came to be regarded as perhaps the greatest experimental scientist of all time. The first unambiguous evidence for the existence of ions, even in the solid state, was obtained by Sir Lawrence Bragg in the course of his X-ray diffraction studies of sodium chloride.

The first error is the assumption that materials do not exist as ions even before they dissolve. The X-ray diffraction of crystalline solids, as we saw in Chapter 5, shows that many solids are collections of ions. There is clearly no force in the argument that 'NaCl molecules' are unlikely to fall apart in water if solid sodium chloride is itself a collection of Na^+ and Cl^- ions.

The second error is to suppose that there are no energy advantages arising from going into solution as ions. For instance, we know (from X-ray studies) that solid benzoic acid is not ionic, yet when it dissolves in water it falls apart (at least partially) into ions. There must therefore be another contribution to the energy that is strong enough to overcome both the forces that hold the molecules together in the solid and the force that keeps the H^+ ion attached to its parent molecule.

We shall find it useful to use the following language. An *ion* is an electrically charged species. A *positively-charged ion* is called a *cation*; a *negatively-charged ion* is called an *anion*. An *electrolyte* is a species capable of providing ions in solution. A *strong electrolyte* is a substance (such as sodium chloride or hydrogen chloride) that is completely dissociated into ions in solution. Whether or not the species is ionic in the *un*dissolved form is irrelevant (for example, sodium chloride is ionic in the solid, while hydrogen chloride is a gas of discrete covalent molecules, yet both are found to be strong electrolytes in water). A *weak electrolyte* is a substance that is only slightly dissociated in solution (such as ethanoic acid in water).

Dissolving

If an ionic solid is to dissolve there must be energetically favourable interactions between the solvent and the dissolved ions to compensate for

8.1 The hydration of ions

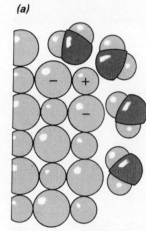

(a)

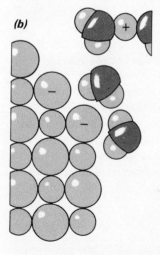

(b)

8.2 Dissolving in water

losing the favourable interactions in the ionic lattice. This new favourable interaction arises from the presence of the dipoles (Section 2.2) of the solvent molecules: the positive end of the dipole can interact favourably with anions, and the negative end with cations. This is illustrated in the case of water as solvent in Fig. 8.1. The grouping of solvent molecules around ions is called *solvation* in general, and *hydration* when the solvent is water.

Dissolving can be pictured as in Fig. 8.2. The water molecules nibble at the edge of the crystal, and the removal of the ions is aided by the favourable energy change that accompanies each act of hydration. Dissolution is the continuation of this process of dislodging ions from the lattice under the influence of the electrostatic interactions with the dipole moments of the polar solvent molecules. This description accounts qualitatively (at least) for the very low solubility of calcium carbonate (and many other carbonates and hydroxides) in water, because the ions are so strongly bound that the hydration energy is insufficient to compensate for the loss of the binding energy. It also accounts for the insolubility of ionic species in non-polar solvents, for they have no dipole and the solvation energy, arising only from van der Waals forces, is very small.

Although qualitatively correct, the picture is incomplete. We can see this by examining the energy changes that occur when sodium chloride dissolves in water. It turns out that sodium chloride (like many other salts) needs energy to go into solution (in the language of thermodynamics, Chapter 9, the dissolution is *endothermic* to the extent of about $4 \, kJ \, mol^{-1}$). In other words, instead of the dissolution of the solid being accompanied by a *decrease* in energy, there is an *increase*! In practice this means that the solution cools slightly when salt dissolves in water, because the energy requirement is taken from the motion of the water molecules and from the walls of the vessel.

Why, then, does salt dissolve? Why does it dissolve if, overall, there appears to be an energy *disadvantage* in so doing? The explanation is provided by thermodynamics (Chapter 11), but the qualitative reason is easy enough to understand. We have stressed already (Section 7.2) that systems have a natural tendency to become chaotic. A crystal is an ordered collection of ions; a solution is a random collection. If there were no forces holding the ions together, they would tend to drift off into any solvent in contact with them (like a gas expanding into a vacuum). The natural direction of change is dissolution, because it corresponds to an increase in chaos. The forces between the ions oppose this tendency. In the presence of a polar solvent, however, the effectiveness of their opposition is reduced because there is now very little energy difference between the crystal and the solution. In many cases the tendency to chaos succeeds in spreading the ions through the solution. In the case of sodium chloride in water (and of other materials that dissolve with a small increase in energy) the tendency to disperse through the solvent overcomes the small energy disadvantage. In contrast, in the case of sodium chloride in benzene, or calcium carbonate in water, the tendency to chaos is present, but unable to lead to dispersal because the energy change opposes it too strongly.

Principles of physical chemistry

8.2 Properties of ionic solutions

The direct evidence for the existence of ions in solids (and, by implication, in solution too) comes from X-ray crystallography. The indirect evidence for their existence in solution comes from various observations, including electrolysis, electrical conduction, and the measurement of colligative properties.

Electrolysis

The general term for chemical change brought about by an electric current is *electrolysis* (lysis means 'breaking'). It may take the form of deposition of metal from a solution, evolution of gas, and so on. The earliest quantitative experimental observations were made by Faraday, and his conclusions are summarized in two laws:

Faraday's first law: The mass of a substance liberated at an electrode is proportional to the charge passed.

Charge is measured by noting the current (I) and the time (t) for which it is passed; then charge $= It$. With current in amperes, A, and time in seconds, s, charge is expressed in coulombs, C, because $1 \, A \, s = 1 \, C$. A stated quantity of charge (*e.g.*, so many coulombs) corresponds to a definite number of electrons because each electron carries a charge of magnitude $1.602 \times 10^{-19} \, C$.[1] Therefore the law can be expressed with 'number of electrons' in place of 'charge'. The law is then easily explained on the basis that electrolysis depends on the supply of electrons to charged species in solution.

Faraday's second law: The amount of substance liberated is a whole-number fraction of the amount of electrons passed.

(This is a modern version of the law.) That is, in order to liberate 1 mol of substance (*e.g.*, 1 mol Zn), a whole number of moles of electrons (*e.g.*, 2 mol electrons) has to be passed. This gives an easy way of determining the charge type of the ions shown to exist by the first law. We can conclude, for instance, that zinc ions exist in solution as Zn^{2+}. Likewise silver ions exist as Ag^+, as shown by the following *Example*.

[1] The magnitude of the charge carried by unit amount of electrons is Le. This quantity occurs widely throughout electrochemistry and is called the *Faraday constant, F*:

$$F = (1.6022 \times 10^{-19} \, C) \times (6.022 \times 10^{23} \, mol^{-1}) = 9.648 \times 10^4 \, C \, mol^{-1}.$$

Example: Measurements of the masses of metals deposited by electrolysis gave the following results. When a current of 1.47 A was passed through an aqueous solution of copper(II) sulphate for 525 s, 254 mg of copper was deposited. The same current passed for the same time through aqueous silver nitrate led to the deposition of 863 mg of silver. What are the charge types of the copper and silver ions in the solutions?

Method: Determine the charge passed (current $\times$ time) and express it as an amount of electrons by noting that the magnitude of the charge on 1 mol of electrons is the Faraday constant. Express the masses of the deposited metals as amounts of substance, and note that a univalent ion is discharged by one electron and a divalent ion by two.

Answer: The charge passed through each solution is $(1.47\,A) \times (525\,s) = 772\,A\,s = 772\,C$ (since $1\,A\,s = 1\,C$). The amount of electrons passed is therefore

$$n(e^-) = \frac{\text{charge}}{F} = \frac{772\,C}{9.648 \times 10^4\,C\,mol^{-1}} = 8.00 \times 10^{-3}\,mol.$$

The amount of substance of Cu deposited is

$$n(Cu) = \frac{254 \times 10^{-3}\,g}{63.55\,g\,mol^{-1}} = 4.00 \times 10^{-3}\,mol.$$

Therefore, one atom is deposited for every two electrons passed, and so the charge type of copper ions is Cu^{2+}. The amount of substance of Ag deposited is

$$n(Ag) = \frac{863 \times 10^{-3}\,g}{107.9\,g\,mol^{-1}} = 8.00 \times 10^{-3}\,mol.$$

Therefore, one atom of Ag is deposited for every electron passed, and so the charge type of the ions present in the solution is Ag^+.

Comment: The silver *coulometer* can be used to measure the charge passed through a circuit. It is basically an electrolysis cell. The mass of silver deposited is measured and interpreted as an amount of electrons passed using a calculation of the type set out in this *Example*, but with the charge type, Ag^+, known.

Colligative properties

Colligative properties (Chapter 7) depend on the *number* of solute particles present in solution and not on their nature. In the case of a solution of sodium chloride, the osmotic pressure, the depression of freezing temperature, and the elevation of boiling temperature are all twice as great as would be expected on the basis that the solute dissolved as 'NaCl molecules'. This is readily accounted for on the basis that the species present are Na^+ and Cl^-, because then there are twice as many solute particles.

The usual way of expressing the increase in the magnitude of the colligative properties is in terms of the *van't Hoff i-factor*. If m is the solute molality we write:

$$
\left.
\begin{aligned}
&\textit{Elevation of boiling temperature}: \quad \delta T = iK_b m \\
&\textit{Depression of freezing temperature}: \delta T = iK_f m \\
&\textit{Osmotic pressure}: \qquad\qquad\qquad \Pi V = inRT.
\end{aligned}
\right\} \tag{8.2.1}
$$

For dilute solutions i can be interpreted as the number of ions each formula unit produces in solution (*e.g.*, $i = 2$ for NaCl, $i = 3$ for Na_2SO_4). An example will make this clear. Consider the effect of nitric acid on the freezing temperature of sulphuric acid. This potent mixture is used in the laboratory for the nitration of aromatic compounds. When the freezing temperature is measured for a given concentration of nitric acid in sulphuric acid, four times the expected depression is observed, hence $i = 4$. This result points to the dissociation of the nitric acid, and the factor 4 can be explained on the assumption that the nitric acid reacts as follows:

$$HNO_3 + 2H_2SO_4 \rightarrow H_3O^+ + NO_2^+ + 2HSO_4^-,$$

there now being four particles present for every HNO_3 molecule added. This observation was used as evidence for the role of the NO_2^+ ion in the nitration reaction.

Example: Common salt is spread on roads to prevent the formation of ice. What depression of freezing temperature of water results from the addition of 100 kg salt to 1 tonne of water?

Method: Use eqn (8.2.1). The cryoscopic constant is given in Table 7.1 as $1.86 \; K/(mol \; kg^{-1})$; the van't Hoff i-factor for sodium chloride is approximately 2 as Na^+ and Cl^- ions go into solution for every NaCl unit added. Calculate the molality of the solution, m, on the basis that $M_r(NaCl) = 22.99 + 35.45 = 58.44$ (see inside back cover).

Answer: The molality of the solution is

$$m = \frac{(100 \times 10^3 \; g)/(58.44 \; g \; mol^{-1})}{1000 \; kg} = 1.71 \; mol \; kg^{-1}.$$

The depression of freezing temperature is therefore

$$\delta T = iK_f m = 2 \times [1.86 \; K/(mol \; kg^{-1})] \times 1.71 \; mol \; kg^{-1} = 6.36 \; K.$$

Comment: The solution will therefore not freeze until the temperature has fallen below $-6\,°C$. Note that the molality of the solution is quite high, and so it will not be ideal. Nevertheless the measured freezing temperature depression ($-6.4\,°C$) is almost exactly what this calculation predicts. The i-factor is not exactly 2 because ion–ion interactions are significant in solution.

8.3 The conductivities of ions

In metals the cations are in a rigid array and charge is transported by the mobile electrons. In electrolyte solutions both anions and cations are free to move, and so both transport charge. The ability of a solution of an electrolyte to carry current is therefore determined by the numbers of anions and cations present and how readily they move (their *mobilities*).

8.3 A conductivity cell

The measurement of conductivity

The resistance, R, of a conductor increases with its length, l, but decreases with its cross-sectional area, A. Therefore we write

$$R \propto l/A \quad \text{or} \quad R = \rho l/A. \tag{8.3.1}$$

ρ (rho) is called the *resistivity* of the substance. The *conductivity* (κ, kappa) is defined as the reciprocal of the resistivity:

$$\text{Conductivity: } \kappa = 1/\rho = l/RA. \tag{8.3.2}$$

Resistance is normally expressed in *ohms* (Ω, omega). The dimensions of conductivity are length/(resistance × area), which is normally expressed as $cm/(\Omega \times cm^2)$, or $\Omega^{-1}\,cm^{-1}$. The internationally recommended unit for Ω^{-1} is the *siemens*, S; when S is used, a conductivity is expressed as so many $S\,cm^{-1}$.

The resistance of an electrolyte solution is measured by connecting a sample cell, like the one illustrated in Fig. 8.3, to one arm of a bridge (a network of resistances, Fig. 8.4). In practice, an *a.c. bridge* has to be used so as to avoid the electrolysis that would occur if current flowed continuously in a single direction. The conductivity of the solution is obtained from the measured resistance by forming $\kappa = l/RA$. In practice it proves to be more accurate to calibrate the cell using a solution of known conductivity, and then to measure the unknown conductivity by comparison.

Molar conductivity

The conductivity of a sample depends on the concentration of the ions. The normal practice is to express the conductivity as a *molar conductivity*, Λ_m (Λ is lambda), by dividing the measured conductivity by the concentration of *added* electrolyte, c:

$$\text{Molar conductivity: } \Lambda_m = \kappa/c. \tag{8.3.3}$$

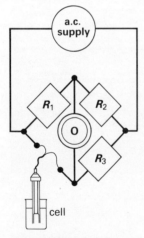

8.4 A bridge circuit for measuring conductance. (R_1, R_2, and R_3 are variable resistances, and O is an oscilloscope)

Example: The conductivity of an aqueous solution of potassium chloride at a concentration $0.10\,mol\,dm^{-3}$ is $0.0129\,S\,cm^{-1}$ at $25\,°C$. What is its molar conductivity?

Method: Use eqn (8.3.3). The units may be manipulated on the basis that $1\,dm^3 = 10^3\,cm^3$.

Answer: From eqn (8.3.3) with $\kappa = 0.0129\,S\,cm^{-1}$ and $c = 0.10\,mol\,dm^{-3}$,

$$\Lambda_m = \kappa/c = \frac{0.0129\,S\,cm^{-1}}{0.10\,mol\,dm^{-3}} = 0.129\,S\,cm^{-1}/(mol\,dm^{-3}).$$

$$= 0.129\,S\,dm^3\,cm^{-1}\,mol^{-1} = (0.129\,S)(1000\,cm^3)\,cm^{-1}\,mol^{-1}$$

$$= 129\,S\,cm^2\,mol^{-1}.$$

Principles of physical chemistry

Strong electrolytes have molar conductivities that are found to be almost independent of the concentration, at least at low concentrations. This is illustrated in Fig. 8.5 for solutions of various chlorides. The measured molar conductivities decrease slightly with increasing concentration of added salt, but there is no dramatic change. So as to be precise, we tabulate the *molar conductivity at infinite dilution*. This is denoted Λ^{∞} and is marked on the diagram.

The molar conductivities of weak electrolytes, however, show a marked dependence on the concentration (this lets us identify them). Figure 8.5 shows that the molar conductivity is quite low at moderate concentrations, remains low as the concentration is reduced, and, when the concentration is extremely low, rises to values similar to the molar conductivities of strong electrolytes. The reason for this behaviour is that the proportion of dissociated molecules of weak electrolytes is low at ordinary concentrations, but increases steeply at very low concentration (high dilution). This behaviour is normally expressed in terms of the *degree of dissociation*, α, defined so that for an electrolyte AB (*e.g.*, CH$_3$COOH)

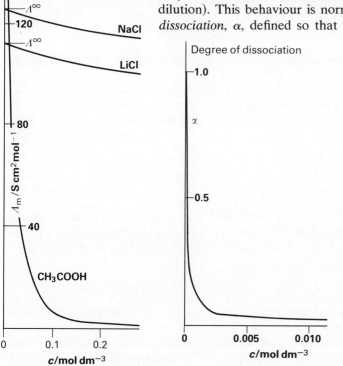

8.5 The molar conductivities of weak and strong electrolytes

8.6 The degree of dissociation and the concentration of weak electrolyte (ethanoic acid in water at 25 °C)

added at a concentration c the concentrations of A^+ [e.g., $H^+(aq)$] and of B^- (e.g., CH_3COO^-) are each αc [and the concentration of AB itself in solution is $(1-\alpha)c$]. The value of α depends on c itself, and its calculation is described in the Appendix, p. 125. All we need at this stage is the form of the result, and the degree of dissociation is plotted in Fig. 8.6. The degree of dissociation is close to unity (complete dissociation) at very low concentrations of added electrolyte, but at higher ('typical') added concentrations the degree of dissociation is very low. This is exactly the kind of behaviour needed in order to explain the conductivities of solutions of weak electrolytes. This point will be developed after another feature has been introduced.

The contributions of individual ions

Extensive studies of the conductivities of electrolytes were made by Kohlrausch. He found that a given type of ion makes a characteristic contribution to the molar conductivity of an electrolyte.

> *Kohlrausch's law: At infinite dilution, the molar conductivity is the sum of the molar conductivities of the ions present.*

The mathematical expression of this law is

> *Kohlrausch's law:* $\Lambda^\infty(AB) = \Lambda^\infty(A^+) + \Lambda^\infty(B^-)$, $\qquad$ (8.3.4)

where the $\Lambda^\infty(A^+)$ and $\Lambda^\infty(B^-)$ are the individual ion molar conductivities (at infinite dilution). Some values are listed in Table 8.1.

Table 8.1 Molar conductivities of individual ions in water at infinite dilution and 25 °C, $\Lambda^\infty/S\,cm^2\,mol^{-1}$

H^+	350	Mg^{2+}	106	Al^{3+}	189	OH^-	199	NO_3^-	71.4
Li^+	38.7	Ca^{2+}	119			F^-	55.4	CH_3COO^-	40.9
Na^+	50.1	Ba^{2+}	127			Cl^-	76.4	SO_4^{2-}	160
K^+	73.5	Cu^{2+}	107			Br^-	78.1		
Ag^+	61.9					I^-	76.8		

The data in Table 8.1, together with Kohlrausch's law, make it possible to predict the molar conductivity of a dilute solution of a strong electrolyte. For instance, the molar conductivity of sodium sulphate solution (at infinite dilution) is

$$\Lambda^\infty(Na_2SO_4) = 2\Lambda^\infty(Na^+) + \Lambda^\infty(SO_4^{2-}) = 260.2\ S\,cm^2\,mol^{-1}.$$

The law also accounts for the concentration dependence of the molar conductivities of solutions of weak electrolytes. Consider the AB species treated in the preceding section. The conductivity when the added electrolyte concentration is c is the sum of the conductivities due to the A^+ ions (at a concentration αc) and the B^- ions (also at a concentration αc). In other words, the total conductivity is

$$\kappa = \alpha c\,\Lambda(A^+) + \alpha c\,\Lambda(B^-) = \alpha c\{\Lambda(A^+) + \Lambda(B^-)\}.$$

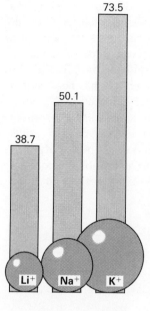

8.7 Ionic molar conductivities and radii

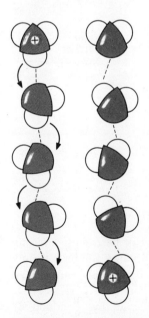

8.8 The mechanism of conduction by hydrogen ions in water

Since there is only a small degree of dissociation, the individual ion conductivities can be replaced by their values at infinite dilution, Λ^∞. It follows that the molar conductivity, κ/c, is

$$\Lambda_m \approx \alpha\{\Lambda^\infty(A^+) + \Lambda^\infty(B^-)\}. \tag{8.3.5}$$

In other words, the molar conductivity is proportional to the degree of dissociation, and therefore rises sharply from a low value when c is large (α much less than 1) to a high value when c is very small (and $\alpha \approx 1$), just as in Fig. 8.5. Note that since the last equation can be written as $\Lambda_m \approx \alpha\Lambda^\infty$, and Λ^∞ can be obtained from tables, the degree of dissociation can be determined by measuring Λ_m.

Ion mobilities

Casual thinking about the relation between the sizes of ions and their molar conductivities would probably suggest that small ions should be more mobile and therefore have greater molar conductivities. A glance at Table 8.1 shows that the opposite is true. Figure 8.7 shows how molar conductivities vary with ionic radii (obtained from X-ray studies of crystals).

The missing feature is the effect of solvation. When an ion moves through water it carries its *hydration sphere*, the water molecules attached to it, along with it, and so its *effective* size may be much larger than the size of the ion itself. A small ion gives rise to a higher electric field at its surface than a big one of the same charge, and so grips its hydration sphere more strongly. The outcome is that a small ion like Li^+ has a *greater* effective size than has K^+ and consequently is less mobile. The increase in the molar conductivities of ions on going down a given group of the periodic table can therefore be explained in terms of their decreasing effective sizes when the effects of hydration are taken into account.

There is, however, a glaring discrepancy. The hydrogen ion, H^+, is extremely small and can therefore be expected to be strongly hydrated, which is why it is normally denoted H_3O^+ or $H^+(aq)$. It is therefore expected to have a very low molar conductivity. Its actual molar conductivity is in fact the highest for any ion. The explanation lies in the fact that H^+ conducts by an entirely different mechanism. It is illustrated in Fig. 8.8. There is only an *effective* transfer of protons through the solution, and the charge of the proton at the top in Fig. 8.8 is carried to the other end of the chain of water molecules by a small motion of the bonds and hydrogen bonds. This gives the same effect as if the proton actually moves, but it takes place much more quickly. Similar remarks apply to the OH^- ion, which also has a high molar conductivity.

Applications of conductivity measurements

Conductivity measurements are used to find concentrations of ions, such as in water-softening equipment. They are also used to distinguish between weak and strong electrolytes, to measure the degree of dissociation, to determine solubilities of sparingly soluble salts (Chapter 12), and to follow acid–base titrations. The apparatus for a *conductimetric titration* consists of a conductivity cell dipping into the beaker where the titration is taking place. The important point to remember throughout the following discussion is that the molar conductivities of $H^+(aq)$ and $OH^-(aq)$ are very high.

Consider the changes of conductivity when a strong acid (such as hydrochloric acid) is titrated against a strong alkali (such as aqueous sodium hydroxide). Initially the solution contains Na^+ and OH^- ions, and so the conductivity is high. As the titration proceeds some of the OH^- ions are replaced by Cl^- ions. As a result, the conductivity falls sharply, Fig. 8.9(a). At the *end-point* (when exactly enough acid has been added to neutralize the alkali) all the OH^- ions have been replaced by Cl^- ions, and the solution has a conductivity typical of a solution of sodium chloride in water. After the end-point, the further addition of acid results in the presence of $H^+(aq)$ ions in the solution, and the conductivity rises sharply. The end-point is indicated by the clear minimum of the conductivity plot. This titration technique is useful if the solutions are coloured and indicators cannot be used.

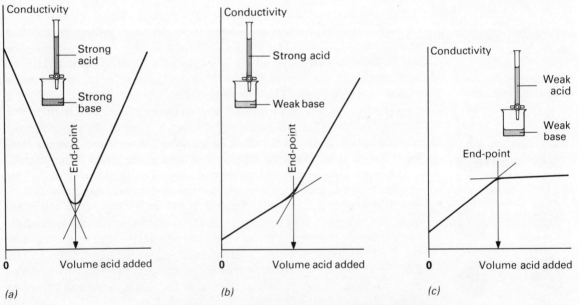

8.9 Conductimetric titrations: (a) strong acid/strong base, (b) strong acid/weak base, (c) weak acid/weak base

Now consider the case of the titration of a weak base with a strong acid, Fig. 8.9(b). Initially the conductivity is low because the base is only slightly dissociated. Up to the end-point the addition of acid results in the formation of the fully dissociated salt, and so the conductivity rises. After the end-point, when $H^+(aq)$ ions become available, the conductivity rises sharply.

Conductimetry is one of the principal ways of detecting the end-points of weak acid/weak base titrations (for example, for ethanoic acid and

Principles of physical chemistry

aqueous ammonia titrations). Initially there is only weakly dissociated base. As weak acid is added, the small number of OH^- ions is replaced, and the conductivity rises as salt (a strong electrolyte) is formed, Fig. 8.9(c). After the end-point, the conductivity barely changes because the addition of the weakly dissociated acid does not result in a significant increase in the number of conducting ions present. The pronounced change in slope at the end-point makes it quite easy to locate.

Appendix: The degree of dissociation

The calculation of α depends on knowing that the equilibrium $AB \rightleftharpoons A^+(aq) + B^-(aq)$ is characterized by the constant

$$K_c = \{[A^+][B^-]/[AB]\}_{eq}.$$

This concept is discussed in Chapter 10. In the present case $[A^+] = \alpha c$, $[B^-] = \alpha c$, and $[AB] = (1-\alpha)c$; therefore

$$K_c = \frac{(\alpha c)(\alpha c)}{(1-\alpha)c} = \frac{\alpha^2 c}{1-\alpha}.$$

This rearranges into

$$(1-\alpha)K_c = \alpha^2 c \quad \text{or} \quad \alpha^2 + (K_c/c)\alpha - (K_c/c) = 0,$$

a quadratic equation with the (acceptable) root

$$\alpha = (K_c/2c)\{-1 + \sqrt{[1 + (4c/K_c)]}\},$$

which is plotted in Fig. 8.6.

Summary

1   An *electrolyte* is a substance capable of supplying ions.
2   A *strong electrolyte* is fully dissociated in solution; a *weak electrolyte* is only slightly dissociated in solution (except at extreme dilutions). The classification depends on the nature of the solvent.
3   A *cation* is a positively-charged ion; an *anion* is a negatively-charged ion.
4   *Solvation* (*hydration* when water is the solvent) is the surrounding of a species in solution by a cluster of solvent molecules.
5   Many dissolutions are accompanied by an increase in energy (are endothermic). The driving force for dissolution is the natural tendency towards chaos of ordered structures; this is successful if only a little energy has to be supplied to the system from the surroundings.
6   *Electrolysis* is chemical change brought about by the passage of an electric current.
7   *Faraday's first law* is *mass liberated is proportional to charge passed*. It indicates the presence of ions in the solution.
8   *Faraday's second law* is *amount of substance liberated is a whole-number multiple of the amount of electrons passed*. It provides a way of determining the charge type of ions in solution.

9   The *Faraday constant* is the magnitude of the charge carried per unit amount of electrons, $F = eL$.

10   In very dilute solutions the *van't Hoff i-factor*, eqn (8.2.1), can be interpreted as the number of ions into which each formula unit dissociates.

11   The *colligative properties* of ionic solutions show that ions exist in solution because they are proportional to the number of particles.

12   The principal *evidence* for the existence of ions comes from X-ray crystallography, the observation that solutions of electrolytes conduct electricity, Faraday's laws, and colligative properties.

13   *Conductivity* (κ) is related to resistance R through $\kappa = l/RA$, where l is the length of the sample, and A its cross-section. The conductivity is normally measured with the cell acting as the unknown resistance in one arm of an a.c. bridge.

14   The *units* of conductivity are $S\,cm^{-1}$ (S is siemens, $1\,S \equiv 1\,\Omega^{-1}$, Ω is ohm).

15   The molar conductivity is denoted Λ_m and defined as κ/c, where c is the concentration of electrolyte added. Its units are $S\,cm^2\,mol^{-1}$; Λ^{∞} is the *molar conductivity at infinite dilution*.

16   In the case of strong electrolytes, the molar conductivity is only slightly dependent on concentration (for low concentrations; high dilutions).

17   In the case of *weak electrolytes*, as the concentration decreases the molar conductivity increases sharply from a low value to a value typical of strong electrolytes.

18   The behaviour of weak electrolytes is expressed in terms of the *degree of dissociation* α, which is very small except at very low concentrations, when it rises to unity.

19   *Kohlrausch's law* states that at infinite dilution the molar conductivity can be expressed as the sum of the molar conductivities of the ions that are present; $\Lambda^{\infty} = \Lambda^{\infty}(A^+) + \Lambda^{\infty}(B^-)$ for an A^+B^- electrolyte.

20   The degree of dissociation of a weak electrolyte can be found from $\alpha = \Lambda_m/\Lambda^{\infty}$.

21   Small ions are more strongly solvated than are large ions; as a consequence their *effective sizes* are larger and their conductivities lower; $H^+(aq)$ and $OH^-(aq)$ ions have unusually high conductivities.

22   Conductivity measurements are used in *conductimetric titrations* and to determine the concentrations of ions.

Problems

1   Describe the evidence for the existence of ions in solution.

2   Account for the following facts: (*a*) many ionic compounds are readily soluble in water, (*b*) iodine is more soluble in tetrachloromethane than in water, (*c*) all nitrates are soluble in water but most hydroxides and carbonates are insoluble.

3   How much charge is transported when a current of 2 A flows through an electric light for 30 s? How may electrons are transported? What amount of substance of electrons (*i.e.*, how many moles of e^-) does that correspond to?

4   For how long would a 100 A current need to flow through an industrial

electrolytic cell in order to generate (a) 1 mol Cl_2 from molten sodium chloride, (b) 3 mol Cu from copper(II) sulphate solution?

5   Which would bring about the greater depression of freezing point, (a) 0.20 g sodium chloride, (b) 0.22 g calcium chloride, when dissolved in the same volume of water?

6   Define the terms *resistance, resistivity, conductivity*, and *molar conductivity*.

7   Identify the errors in the following remarks: (a) the conductivity of sodium chloride in water increases as its concentration decreases, (b) the molar conductivity of sodium chloride in water is independent of concentration.

8   Estimate the molar conductivities of dilute solutions of (a) copper(II) sulphate, (b) potassium nitrate. What will be the resistance of a solution of cross-section 1 cm^2, length 10 cm of 0.01 mol dm^{-3} potassium nitrate solution?

9   The molar conductivities at infinite dilution of three of the alkali metal halides were measured and found to be as follows: Λ^{∞}(LiCl) = 115, Λ^{∞}(NaCl) = 126, Λ^{∞}(KCl) = 150, all in S cm^2 mol^{-1}. Account for the trend in these values. Would you expect the molar conductivities of the bromides to be uniformly larger or smaller? Confirm your conclusion on the basis of the data in Table 8.1.

10   The molar conductivity of 0.05 mol dm^{-3} aqueous ethanoic acid is 7.36 S cm^2 mol^{-1}. What is its degree of dissociation at this concentration?

11   (a) Describe briefly an experiment you could carry out to illustrate the relationship between the charge on an ion in solution and the quantity of electrical charge required for its discharge.

(b) A current of 0.5 A was passed through 200 cm^3 of an aqueous solution of silver nitrate of concentration 0.1 mol dm^{-3} for one hour. The anode was platinum and the cathode was silver.

(i) Give the equation for the discharge of the metal ion.

(ii) What mass of silver would be deposited in this time?

(iii) What is the concentration in mol dm^{-3} of the solution left with respect to both silver nitrate and nitric acid?

(iv) What gas would be discharged at the anode? What changes would occur if the platinum anode was replaced with a silver anode?

(*AEB 1980 part question*)

12   (a) Describe with essential detail the experiments you would carry out to obtain a value for the molar conductivity at infinite dilution [Λ^{∞}] of hydrochloric acid in aqueous solution at 25 °C.

(b) State what information you would need, and explain how you would use it, to evaluate [Λ^{∞}] for ethanoic acid (acetic acid) in aqueous solution at 25 °C. [Details of how the necessary data are obtained are not required.]

(c) At 25 °C, [Λ^{∞}] for ethanoic acid is 391 ohm^{-1} cm^2 mol^{-1} [S cm^2 mol^{-1}], and the dissociation constant of this acid is 1.76×10^{-5} mol dm^{-3}. Calculate to two significant figures the electrolytic conductivity (specific conductance) of a 0.100 molar [mol dm^{-3}] solution of ethanoic acid at 25 °C.

(*Oxford*)

9

Energy in transition: thermochemistry

In this chapter we investigate the role of energy in chemical reactions. We meet the First law of thermodynamics, and see some of its applications to chemistry. We see how to measure energy changes occurring in the course of reactions, and how to manipulate thermochemical data so that we can make predictions about the energy they release or require. The kind of information described in this chapter lies at the heart of discussions of the properties of fuels, of industrial syntheses, and of the energies involved in the activity of biological cells.

Introduction

We are often interested in transfers of energy and its change from one form to another. For instance, burning a fuel in a rocket changes the energy stored in chemical bonds into the kinetic energy of the space vehicle. Burning gas, oil, or coal can be used as a source of heat, or it may be used to generate electricity, which in turn enables energy to be transported over great distances and used to provide heat or to do work elsewhere. The chemist needs some way of assessing how much energy can be released in a reaction, or how much a reaction needs for it to take place; *thermochemistry* provides the necessary information.

Thermochemistry is a part of a much more general subject known as *thermodynamics*, the science of the transformations of energy. This is the subject developed in this and the next four chapters. It grew out of the study of heat engines (in particular, of steam engines) when there was intense interest, during the nineteenth century, about the way their efficiencies could be improved. These origins, however, have been left behind, and modern thermodynamics has become a subject of extraordinary power and wide application, Box 9.1.

Box 9.1: The development of thermodynamics

Thermodynamics grew out of the industrial revolution, when engineers and scientists turned their attention to the improvement of the efficiencies of steam engines. Sadi Carnot (b. 1796 in Paris) was convinced that France's military defeat was due to the inferiority of her steam power, and investigated ways of improving it. This led to the formulation of the concept of *entropy* (Chapter 11). J. P. Joule (b. Salford, 1818) studied under Dalton, and then turned to a precise study of the properties of heat and work and their interrelation. He was able to demonstrate their interconvertibility, and hence the concept of *energy* entered physical science. Rudolph Clausius (mentioned in Chapter 8) is generally given the credit for turning the subject into an exact science by sharpening the definitions of the quantities involved and showing how to manipulate them mathematically. One of the giants of the subject is William Thomson, later Lord Kelvin (b. 1824 in Belfast). He was a professor at the University of Glasgow at 22, and remained there throughout his life. In the course of that life he became wealthy and famous largely through his inventions relating to telegraphy and submarine cables, but his enduring fame comes from his contribution to thermodynamics, the formulation of the First law of thermodynamics, and the conservation of energy. His name is commemorated in the temperature scale, for another major contribution that he made to thermodynamics was the exact definition of the 'temperature' of a body. The most important contributor to the application of thermodynamics to chemical problems (it had emerged out of engineering and moved into physics) was Josiah Willard Gibbs. He was born in Connecticut in 1839 and was described by a contemporary historian as 'the greatest of Americans, judged by his rank in science'. The man who formulated the connection between the properties of individual molecules and the bulk properties of samples, the branch of physical chemistry known as *statistical thermodynamics*, was Ludwig Boltzmann (b. 1844 in Vienna) who committed suicide, ill and depressed, in 1906.

9.1 The First law

The First law of thermodynamics is a summary of a lot of experience. (This is a common feature of how scientific laws are formulated.) In particular, no one has ever succeeded in constructing a perpetual motion machine, one that will generate work without consuming fuel. The hope has been given up, and the failure to generate energy out of nothing is now expressed as a law:

> *First law of thermodynamics: Energy can be neither created nor destroyed.*

(Now that we are familiar with nuclear energy, we have to interpret 'energy' as including mass through $E = mc^2$.)

Work and heat

When interpreting the law several points have to be kept in mind about the ways in which energy may be transferred to and from a system. The *system* means the part of the world we are interested in. In a laboratory that might be a sample in a flask. The rest of the world is called the *surroundings.* Usually that means a water-bath or the atmosphere. One way is by doing *work* on the system. For instance, energy can be transferred to a vessel of water (with the effect of raising its temperature) by stirring it vigorously. It is quite easy to measure the quantity of work done by using an electric motor of known power output or by connecting paddles to a falling weight. For instance, in order to raise the temperature of 500 cm^3 of water from 20 °C to 25 °C an energy transfer of 10.4 kJ is required. This corresponds to a 100 W motor running for 104 s (since 1 J = 1 W s) or letting a 10 kg mass fall through 106 m (since work = *mgh*, g being the acceleration due to gravity, 9.81 m s^{-2}).

Another way of transferring energy is by means of *heat.* For instance, the same rise in temperature of the vessel of water (and therefore the same transfer of energy) can be brought about by immersing a heater in the water, and generating 10.4 kJ of heat (*e.g.*, by running a 100 W heater for 104 s).

Work and heat are the two fundamental ways of transferring energy to or from a system. The system can be thought of as a kind of bank for energy. Transactions with the bank can be made in two kinds of currency: as work or as heat. But once the currencies are inside the bank they are stored simply as its energy reserves, and may be withdrawn in either currency. An *isolated system* is like a bank when it is shut; no work or heat transactions may be made. A stoppered dewar flask is an example.

The quantity of energy transferred as work during a transaction with the system is written *w* (and normally expressed in joules, J, or kilojoules, kJ). So, if we transfer 10 kJ of energy by doing work *on* the system, we write *w* = +10 kJ. If, on the contrary, the system itself *does* 10 kJ of work, we say that *w* = −10 kJ, because the system has *lost* 10 kJ of energy by doing work on the outside world.

The energy transferred as heat is written *q*, and is also normally expressed in joules or kilojoules. So, if 10 kJ of energy is transferred *to* the system as heat, we write *q* = +10 kJ. If the same quantity of energy is transferred as heat *from* the system, we write *q* = −10 kJ, the negative sign indicating loss of energy from the system.

The First law can now be expressed in terms of *w* and *q*. We denote the energy possessed by the system, its *internal energy,* as *U*; the *change* of internal energy when energy is transferred as work or heat is written ΔU. It follows that the change of internal energy is

$$\Delta U = q + w. \tag{9.1.1}$$

Therefore, if 250 kJ of energy is transferred to the system as work (*w* = +250 kJ) and at the same time 150 kJ is transferred back into the environment as heat (*q* = −150 kJ), the overall change in the internal energy is ΔU = +100 kJ; the reserves of the energy bank have risen by 100 kJ as a result of the process, and are stored until a later process withdraws them either as heat or as work.

Principles of physical chemistry

Energy and enthalpy

Suppose energy is transferred as heat to $500\,cm^3$ of water in an open vessel. As well as getting 'hotter' (*i.e.*, reaching a higher temperature) the water expands a little. When the water expands, it pushes back the atmosphere, Fig. 9.1. That pushing means that the system is doing work on its surroundings. In other words, when a quantity of energy, q, is transferred as heat to a system open to the atmosphere, the change in the system's energy is not simply $\Delta U = q$, because it simultaneously loses some energy by pushing back the atmosphere. The quantity that automatically takes this loss into account is the *enthalpy, H*. The enthalpy is defined so that when an energy q is transferred as heat to the system at constant pressure, the change of enthalpy is exactly equal to q:

$$\Delta H = q: \quad q \text{ transferred at constant } pressure. \tag{9.1.2}$$

When the same quantity of energy is transferred to the system under conditions of constant volume, when it can do no work on the outside, there is a change of internal energy of magnitude q:

$$\Delta U = q: \quad q \text{ transferred at constant } volume. \tag{9.1.3}$$

The difference between ΔH and ΔU is illustrated in Fig. 9.2. Usually the

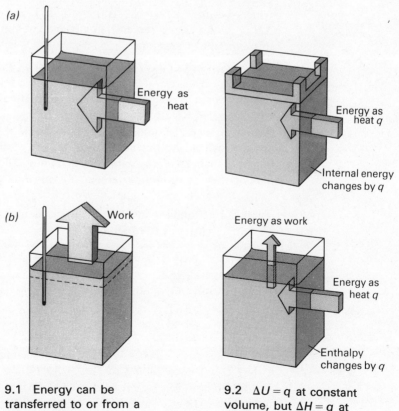

9.1 Energy can be transferred to or from a system as either (a) heat or (b) work

9.2 $\Delta U = q$ at constant volume, but $\Delta H = q$ at constant pressure

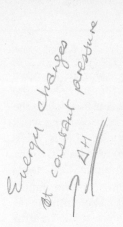

change of volume that occurs when energy is transferred at constant pressure is small, the energy that leaks away as work is therefore also small, and so ΔH and ΔU are almost the same. Reactions that involve gases, however, often lead to big changes of volume, and therefore do a lot of work on the surroundings. For them ΔH and ΔU can be significantly different (several kilojoules per mole of reaction).

In chemistry we are mainly concerned with energy transfers at constant pressure, and so we concentrate on the enthalpy. Once we know the enthalpy changes that occur during reactions (using, for example, the information given in the tables later in this chapter), we can state how much energy that reaction requires or liberates as heat when it takes place at constant pressure.

9.2 Enthalpies of transition

The easiest type of process to consider is a change of phase, such as melting or evaporating. We all know that energy is required in order to evaporate water. In a kettle water boils under conditions of constant pressure (usually a pressure of around 1 atm), and so we can say at once that the energy required as heat in order to vaporize a given amount of water at 1 atm pressure is the enthalpy change in the process *liquid water* → *water vapour* at 100 °C. The *molar enthalpy of vaporization*, $\Delta H_{vap,m}$, of water is the enthalpy change at the specified temperature for unit amount of substance, *e.g.*, 1 mol H_2O, in the process

$$H_2O(l) \rightarrow H_2O(g); \qquad \Delta H_{vap,m} = +40.7 \text{ kJ mol}^{-1}.$$

The numerical value means that in order to vaporize 1 mol $H_2O(l)$, or 18 g of water, 40.7 kJ of energy must be provided. Enthalpies of vaporization of several liquids are listed in Table 9.1. The value for water is anomalously high; this is another aspect of the presence of hydrogen bonds, which have the effect of strapping the molecules together and making vaporization an energetically more demanding process.

Table 9.1 Standard molar enthalpies of melting and vaporization, $\Delta H_m^{\ominus}/\text{kJ mol}^{-1}$

	T_f/K	Melting	T_b/K	Vaporization
H_2	13.96	0.12	20.38	0.92
N_2	63.15	0.72	77.35	5.59
O_2	54.36	0.44	90.18	6.82
H_2O	273.15	6.01	373.15	40.66
				(44.02 at 25 °C)
NH_3	195.40	5.65	239.73	23.35
CH_4	90.68	0.94	111.66	8.18
C_6H_6	278.65	9.83	353.25	30.8

Values relate to the transition temperature

Principles of physical chemistry

Example: Calculate how long it would take for a 2 kW heater (such as in an electric kettle) to evaporate $1 \, dm^3$ of water at $100 \, °C$. The molar enthalpy of vaporization of water at that temperature is $40.66 \, kJ \, mol^{-1}$.

Method: Calculate the total quantity of energy that has to be supplied as heat by forming $n\Delta H_{vap,m}$. To find the amount of substance it is sufficiently accurate for this calculation to take the density of water as $1 \, g \, cm^{-3}$. Calculate the time it takes for a 2 kW heater to supply that quantity of energy from power = (energy)/(time).

Answer: The amount of substance of H_2O to be evaporated is

$$n = \frac{(1000 \, cm^3) \times (1 \, g \, cm^{-3})}{(18.02 \, g \, mol^{-1})} = 55.5 \, mol.$$

The quantity of energy to be supplied at constant pressure is

$$q = n\Delta H_{vap,m} = (55.5 \, mol) \times (40.66 \, kJ \, mol^{-1}) = 2257 \, kJ.$$

The time it takes for this quantity of energy to be supplied by a 2 kW heater (*i.e.*, a heater supplying energy at a rate $2 \times 10^3 \, J \, s^{-1}$) is

$$t = \frac{(2257 \times 10^3 \, J)}{(2 \times 10^3 \, J \, s^{-1})} = 1128 \, s.$$

The kettle will boil to dryness in about 19 minutes.

Comment: The time to heat the water from room temperature to the boiling temperature can be calculated if the heat capacity of the water is known (it is: its value is $75.5 \, J \, K^{-1} \, mol^{-1}$). The calculation could be continued by finding the total time it takes to evaporate water starting at room temperature.

[handwritten in margin: $q = n \, \Delta H_{vap,m}$]

Solids require energy in order to melt. Melting (or *fusion*) normally takes place under conditions of constant pressure, and so the energy absorbed as heat can be identified with the change of the system's enthalpy. We therefore speak of the *enthalpy of melting*, and, for unit amount of substance, of the *molar enthalpy of melting* $\Delta H_{melt,m}$ at the specified temperature. For instance, the molar enthalpy of melting of ice at $0 \, °C$ is the enthalpy change for the process

$$H_2O(s) \rightarrow H_2O(l); \qquad \Delta H_{melt,m} = +6.0 \, kJ \, mol^{-1}.$$

The value signifies that 6.0 kJ are required to melt 1 mol $H_2O(s)$; that is, 18 g of ice. Some other enthalpies of melting are given in Table 9.1.

9.3 Enthalpy in action

By far the most important type of enthalpy change for the chemist is the change that accompanies a chemical reaction. First we deal with the way these enthalpies are measured.

Calorimetry

A calorimeter is a device for measuring the quantity of energy given out or taken in as heat in the course of a reaction. The heat transaction is measured by following the change of temperature it brings about. A change of temperature, δT, can be converted to the energy supplied, q, using the *heat capacity*, C. This is the coefficient C in the expression,

$$q = C\delta T. \tag{9.3.1}$$

For instance, if the heat capacity of a lump of metal is $232\,\text{J K}^{-1}$ (as would be the case for a 600 g block of copper), and during a reaction we observe a temperature rise of 5.2 K, we know that 1.21 kJ of energy has been transferred. The *specific heat capacity*, c, is often listed. This is the heat capacity per unit mass of material. The actual heat capacity of the calorimeter (and its contents) can then be calculated if the masses are known.

There are several ways of putting these ideas into practice. The simplest type of calorimeter is an insulated pot fitted with a thermometer. The reaction is started by mixing the reactants, and the temperature change is recorded. The heat capacity of the system (the calorimeter plus its contents) can be measured in a separate experiment (for instance, by heating it electrically) or calculated from specific heat capacities.

A more elaborate piece of apparatus is the *flame calorimeter* illustrated in Fig. 9.3(*a*). It is used for measuring the enthalpies of combustion of gases and volatile liquids. In a typical experiment, air is passed through the combustion chamber and the energy evolved by the reaction heats the surrounding water-bath. The temperature rise produced by the combustion of a known mass of the compound is measured.

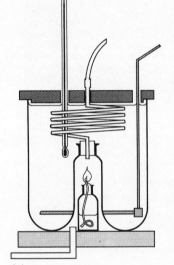

(a)

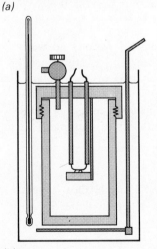

(b)

9.3 (*a*) A flame calorimeter and (*b*) a bomb calorimeter

Example: The combustion of 0.500 g of ethanol in a flame calorimeter produced a temperature rise of 5.01 K. The same rise in temperature was brought about electrically using a 10.0 V supply passing a current of 2.50 A for 600 s (10 min). Find the molar enthalpy of combustion of ethanol.

Method: The quantity of energy supplied by the combustion of ethanol is the same as the quantity supplied in the electrical heating experiment because the temperature rise is the same in each case. Electrical energy is given by (current) × (time) × (potential difference), or ItV for short. Note that $1\,\text{A V s} = 1\,\text{J}$. The molar mass of ethanol is $46.07\,\text{g mol}^{-1}$ (see inside back cover); use this to convert ΔH_c to a molar quantity through $\Delta H_{\text{c.m}} = \Delta H_c/n$.

Answer: The quantity of energy supplied electrically is

$$q = (2.50\,\text{A}) \times (600\,\text{s}) \times (10.0\,\text{V}) = 15\,000\,\text{A V s} = 15.0\,\text{kJ}.$$

The amount of substance of C_2H_5OH in the sample, which also produces this quantity of heat, is

$$n = \frac{0.500\,\text{g}}{46.07\,\text{g mol}^{-1}} = 1.09 \times 10^{-2}\,\text{mol}.$$

　　　Principles of physical chemistry

Therefore, since the enthalpy of combustion of 0.50 g is 15.0 kJ, the molar enthalpy of combustion is

$$\Delta H_{c,m} = \frac{-15.0\,\text{kJ}}{1.09 \times 10^{-2}\,\text{mol}} = -1380\,\text{kJ mol}^{-1}.$$

(ΔH is negative because energy leaves the ethanol as heat.)

Comment: Instead of arranging for the electrical heating to supply the same quantity of energy as the combustion, it is common to determine the heat capacity, C, of the calorimeter from $C = $ (energy as heat)/(temperature rise): in the present case the electrical heating experiment gives $C = 2994\,\text{J K}^{-1}$. Then if in the combustion experiment a temperature rise δT is observed, the heat of combustion is given by $q = C\delta T$. In this way the temperature rises do not have to be matched.

The third basic type is the *bomb calorimeter*, Fig. 9.3(*b*). This is a strong, closed vessel immersed in an insulated water-bath. The sample under investigation is held in the sample holder, and ignited electrically in an atmosphere of high pressure oxygen (for the case of a combustion reaction). The temperature rise of the entire assembly is recorded. Modern calorimeters are so sensitive that very small amounts of material are sufficient. This is especially important when the material is in short supply and expensive, as in the case of biologically important reactions.

The significant point about the use of a bomb calorimeter is that it operates at constant *volume* and so the heat evolution measured is the internal energy change ΔU and not the enthalpy change.

Not all reactions are combustions. For instance, we might be interested in the *enthalpy of solution* of a salt in water (this arises in the discussion of dissolving, Chapter 8). The same type of procedure may be used as outlined above; the salt is added to the water in an insulated vessel, and the change of temperature is recorded and converted to an energy transfer, and then to an enthalpy change. The *enthalpy of neutralization* is a related quantity; it is the enthalpy change that accompanies the neutralization of an acid by a base. Enthalpies of solution and neutralization often refer to the formation of infinitely dilute solutions so as to avoid complications arising from the interactions of ions.

Enthalpies of combustion

One very important type of reaction is combustion, especially the combustion of organic materials to carbon dioxide and water. Information on this type of reaction is central to an understanding of the metabolism of organisms and the utilization of fuels.

The basic reaction is illustrated by the combustion of methane:

$$CH_4(g) + 2O_2(g) \rightarrow CO_2(g) + 2H_2O(l).$$

This reaction takes place when natural gas is burnt, and so this discussion

CH₄(g) + 2O₂(g)

890.4 kJ mol⁻¹

CO₂(g) + 2H₂O(l)

9.4 The enthalpy of combustion of methane

is fundamental to the proper understanding of our energy resources. When the reaction takes place in a flame calorimeter of the kind illustrated in Fig. 9.3(a), the energy evolved as heat (at constant pressure) is 890.4 kJ per mole of CH_4 consumed. The molar enthalpy of combustion of methane is therefore -890.4 kJ mol⁻¹ (the negative sign indicating that energy is released in the combustion).

The last statement has to be made more precise. What is normally reported is a special type of enthalpy change, called the *standard molar enthalpy of combustion*. This refers to the change of enthalpy in a reaction in which all the reactants and products are in their *standard states* at the specified temperature. A standard state is defined as *the most stable state of the substance at a pressure of 1 atm* (101.325 kPa). In the case of methane combustion, the standard states at 25 °C are gaseous methane, oxygen, and carbon dioxide, and liquid water, all at a pressure of 1 atm. If the specified temperature is over 100 °C, the standard states of the species are all gases at 1 atm, because that then is also the stable state of water. The standard enthalpy of a reaction is denoted by a superscript ⊖ on the ΔH. Since reaction enthalpies (and standard states) depend on the temperature, that has to be specified too. For instance, the standard molar enthalpy of combustion of methane at 25 °C, $\Delta H_{c,m}^{\ominus}$ is written

$$CH_4(g) + 2O_2(g) \rightarrow CO_2(g) + 2H_2O(l);$$
$$\Delta H_{c,m}^{\ominus} = -890.4 \text{ kJ mol}^{-1} \quad \text{at } 25\,°C.$$

The negative sign indicates a *release* of energy in the combustion, Fig. 9.4.

Table 9.2 Standard molar enthalpies of combustion at 25 °C, $\Delta H_{c,m}^{\ominus}/\text{kJ mol}^{-1}$

$H_2(g)$	-286	$CH_4(g)$	-890	$C_6H_6(l)$	-3268
C(graphite)	-393.5	$C_2H_6(g)$	-1560	$C_2H_5OH(l)$	-1371
C(diamond)	-395.4	$C_8H_{18}(l)$	-5512	$CH_3CHO(l)$	-1167
S(rhombic)	-296.9	$C_2H_4(g)$	-1411	$CH_3COOH(l)$	-875
S(monoclinic)	-297.2	$C_2H_2(g)$	-1300	$C_6H_{12}O_6(s)$ (glucose)	-2816

Many other combustion reactions are of importance, and some combustion enthalpies are listed in Table 9.2. The enthalpy of combustion of the alkanes increases with the number of CH_2 units, and depends on the degree of chain branching. This is illustrated in Fig. 9.5. The combustion of glucose at 25 °C is the reaction

$$C_6H_{12}O_6(\text{glucose; s}) + 6O_2(g) \rightarrow 6CO_2(g) + 6H_2O(l);$$
$$\Delta H_{c,m}^{\ominus} = -2816 \text{ kJ mol}^{-1},$$

Fig. 9.6, and this immense quantity of energy (which corresponds to 15.6 kJ per gram of glucose) is used by reactions inside living cells. It is the source of energy for animals, which use respiration to tap the energy

Enthalpy of combustion

6000

C_8H_{18}

C_7H_{16}

4000 C_6H_{14}

C_5H_{12}

$-\Delta H_{c,m}^{\ominus}/\text{kJ mol}^{-1}$

C_4H_{10}

2000 C_3H_8

C_2H_6

CH₄

0   2   4   6   8
Number of carbon atoms

9.5 Enthalpies of combustion of some hydrocarbons

Principles of physical chemistry

9.6 The enthalpy of combustion of glucose

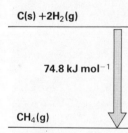

9.7 The enthalpy of formation of methane

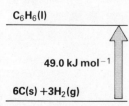

9.8 The enthalpy of formation of benzene

resources provided by digestion and to turn them into metabolic activities.

Enthalpies of formation

The enthalpy of formation of a compound is the change in enthalpy that occurs when it is formed from its elements. When unit amount of a compound is formed from its elements, and every species involved is in its standard state at the temperature specified, the change in enthalpy is called the *standard molar enthalpy of formation* of the compound and is denoted $\Delta H_{f,m}^{\ominus}$.

As an example, consider the standard molar enthalpy of formation of methane at 25 °C:

$$C(s) + 2H_2(g) \rightarrow CH_4(g); \qquad \Delta H_{f,m}^{\ominus} = -74.8 \text{ kJ mol}^{-1}.$$

The standard states at this temperature (25 °C) are graphite, gas, and gas, respectively, all at 1 atm pressure. Note that the standard molar enthalpy of formation of methane is negative, Fig. 9.7, which indicates that 74.8 kJ of energy is *evolved* as heat for each mole of CH_4 formed when the reaction is carried out at 1 atm pressure.

The standard molar enthalpy of formation of benzene is the enthalpy change in the following reaction:

$$6C(s) + 3H_2(g) \rightarrow C_6H_6(l); \qquad \Delta H_{f,m}^{\ominus} = +49.0 \text{ kJ mol}^{-1},$$

Fig. 9.8. This illustrates two points.

The first is that *enthalpies of formation may be either positive or negative*. A positive enthalpy of formation indicates that energy has to be provided in order for the reaction to proceed. In the case of the formation of benzene, 49.0 kJ of energy is needed in order to produce 1 mol C_6H_6 in its standard state at 25 °C.

The second point is that the formation reaction need not be an actual, feasible, reaction. For instance, burning carbon in hydrogen does not in fact generate benzene. Likewise, we can speak of the enthalpy of formation of complex molecules like glucose even though the direct synthesis is

Table 9.3 Standard molar enthalpies of formation at 25 °C, $\Delta H_{f,m}^{\ominus}/\text{kJ mol}^{-1}$

$H_2O(l)$	−285.8	$NH_4Cl(s)$	−314.4	$CH_4(g)$	−74.8
$H_2O_2(l)$	−187.8	$NaCl(s)$	−411.0	$C_2H_2(g)$	226.8
$NH_3(g)$	−46.1	$KCl(s)$	−435.9	$C_2H_4(g)$	52.3
$HF(g)$	−271.1	$KBr(s)$	−392.2	$C_2H_6(g)$	−84.6
$HCl(g)$	−92.3	$KI(s)$	−327.6	$C_6H_6(l)$	49.0
$HBr(g)$	−36.4	$NaOH(s)$	−426.7	$CH_3OH(l)$	−239.0
$HI(g)$	26.5	$Al_2O_3(s)$	−1675.7	$C_2H_5OH(l)$	−277.0
$CO(g)$	−110.5	$SiO_2(s)$	−910.9	$CH_3COOH(l)$	−484.2
$CO_2(g)$	−393.5	$HNO_3(l)$	−173	$CH_3COOC_2H_5(l)$	−486.6
$NO_2(g)$	33.2	$HCl(aq)$	−167.2	$C_6H_{12}O_6(s)$	−1274
$N_2O_4(g)$	9.2	$H_2SO_4(aq)$	−909.3	$C_{12}H_{22}O_{11}(s)$	−2222

impossible. What the enthalpy of formation tells us is the change of enthalpy that *would* occur *if* a reaction pathway could be found that takes one from the elements to the compound of interest. We shall develop this point shortly.

Enthalpies of formation of compounds are very important in thermo-chemistry, as we shall now see. They have been determined (often from enthalpies of combustion, as the following *Example* shows) for many materials over a wide range of temperatures. Some values for 25 °C are listed in Table 9.3.

Example: The standard molar enthalpies of combustion of carbon, hydrogen, and methane are $-393.5\,\text{kJ mol}^{-1}$, $-285.8\,\text{kJ mol}^{-1}$, and $-890.4\,\text{kJ mol}^{-1}$, respectively, at 25 °C. Deduce the value of the standard molar enthalpy of formation of methane at that temperature.

Method: The reaction is $C(s) + 2H_2(g) \rightarrow CH_4(g)$ where the standard state of carbon is graphite (at 25 °C). The three combustion reactions are

$$
\begin{aligned}
&(a) &&C(s) + O_2(g) \rightarrow CO_2(g); &&\Delta H^{\ominus}_{c,m} = -393.5\,\text{kJ mol}^{-1}\\
&(b) &&H_2(g) + \tfrac{1}{2}O_2(g) \rightarrow H_2O(l); &&\Delta H^{\ominus}_{c,m} = -285.8\,\text{kJ mol}^{-1}\\
&(c) &&CH_4(g) + 2O_2(g) \rightarrow CO_2(g) + 2H_2O(l); &&\Delta H^{\ominus}_{c,m} = -890.4\,\text{kJ mol}^{-1}.
\end{aligned}
$$

The reaction of interest can formally be regarded as running as follows: burn $C(s)$; burn $2H_2(g)$; 'unburn' (*i.e.*, run the combustion reaction in reverse) $CO_2(g) + 2H_2O(l)$ to form $CH_4(g)$. The overall result is the same as the required formation reaction. Add the reaction enthalpies for each step.

Answer: The overall enthalpy of reaction is the sum of $(a) + 2(b) - (c)$:

$$
\begin{aligned}
\Delta H^{\ominus}_m &= (-393.5\,\text{kJ mol}^{-1}) + 2 \times (-285.8\,\text{kJ mol}^{-1}) - (-890.4\,\text{kJ mol}^{-1})\\
&= -74.7\,\text{kJ mol}^{-1}.
\end{aligned}
$$

Check that the overall reaction is indeed the formation reaction:

$$
\begin{aligned}
&(a) &&C + O_2 &&= CO_2\\
&2(b) &&2H_2 + O_2 &&= 2H_2O\\
&-(c) &&CO_2 + 2H_2O &&= CH_4 + 2O_2
\end{aligned}
$$

$$
\begin{aligned}
&C + 2H_2 + CO_2 + 2H_2O + 2O_2 = CH_4 + CO_2 + 2H_2O + 2O_2\\
&C + 2H_2 \hspace{4.5cm} = CH_4, \text{ as required.}
\end{aligned}
$$

Comment: Note that this kind of calculation is independent of whether or not the individual reaction steps can actually be carried out in practice; only *formally* possible steps are required. The enthalpy of combustion of methane ($-890.4\,\text{kJ mol}^{-1}$) is high, which is an advantage for its applications in domestic and industrial heating.

Enthalpies of reaction

We now show that the data in Tables 9.2 and 9.3 can be used in the discussion of any reaction, not only combustion and formation, and that by their use we can predict the enthalpy change of any reaction, including reactions that have not been studied experimentally. This is the point where the First law of thermodynamics truly enters chemistry.

Suppose we were interested in some fairly complicated reaction such as

$$HCl(g) + CH_2{=}CH_2(g) \rightarrow CH_3{\cdot}CH_2Cl(g),$$

and we wanted to know the standard molar enthalpy change. The First law, in effect the conservation of energy, tells us that *whatever route we select between reactants and products, the same energy change occurs.*

The path-independence of reaction enthalpies was noticed by Hess in his extensive series of measurements on 'heats of reaction' (which we now call reaction enthalpies) before the thermodynamic basis had been established. In 1840 he summarized his conclusions in a law that now bears his name. In modern language it reads:

Hess's law: The overall reaction enthalpy is the sum of the reaction enthalpies of each step into which the reaction may formally be divided.

As an illustration, consider the $CH_3{\cdot}CH_2Cl$ reaction. We can *formally* regard it (that is, write it on paper, not necessarily actually carry it out in the laboratory) as occurring in the following three steps, Fig. 9.9.

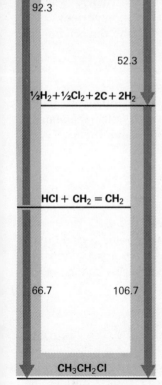

½H₂ + ½Cl₂ + CH₂ = CH₂

92.3

52.3

½H₂+½Cl₂+2C+2H₂

HCl + CH₂ = CH₂

66.7   106.7

CH₃CH₂Cl

(1) $HCl(g) \rightarrow \tfrac{1}{2}H_2(g) + \tfrac{1}{2}Cl_2(g);$ $\Delta H_m^{\ominus} = +92.3 \text{ kJ mol}^{-1}$

(2) $CH_2{=}CH_2 \rightarrow 2C(s) + 2H_2(g);$ $\Delta H_m^{\ominus} = -52.3 \text{ kJ mol}^{-1}.$

(At this stage we have reversed the formation reactions, and so have produced the elements $C(s)$, $H_2(g)$, and $Cl_2(g)$; the enthalpies are taken from Table 9.3 by changing the signs of the enthalpies of formation listed there; they refer to 25 °C.) Now form the elements into the final product:

(3) $2C(s) + \tfrac{5}{2}H_2(g) + \tfrac{1}{2}Cl_2(g) \rightarrow CH_3{\cdot}CH_2Cl(g);$
$\Delta H_m^{\ominus} = -106.7 \text{ kJ mol}^{-1}.$

The enthalpy change is the standard molar enthalpy of formation of $CH_3{\cdot}CH_2Cl$. It follows that the overall enthalpy change is

$$\Delta H_{f,m}^{\ominus} = (+92.3 \text{ kJ mol}^{-1}) + (-52.3 \text{ kJ mol}^{-1}) + (-106.7 \text{ kJ mol}^{-1})$$
$$= -66.7 \text{ kJ mol}^{-1}.$$

According to Hess's law, the same enthalpy change will occur when the actual reaction is carried out under standard conditions, hence the standard molar reaction enthalpy is $-66.7 \text{ kJ mol}^{-1}$.

The classification of reactions

We have seen that some reactions proceed with an increase of enthalpy while others proceed with a decrease. This leads very naturally to a division of reactions into two classes, *exothermic* and *endothermic*:

Exothermic reactions are accompanied by a decrease of enthalpy, $\Delta H < 0$.
Endothermic reactions are accompanied by an increase of enthalpy, $\Delta H > 0$.

9.9 Hess's law and reaction enthalpy. The circuit of reactions is called a *cycle*

All the combustion reactions in Table 9.2 are therefore exothermic reactions, and liberate energy when they take place under standard conditions. Some of the formation enthalpies in Table 9.3 are positive and some are negative. Methane, with a negative enthalpy of formation, is produced by an exothermic formation reaction. It is therefore referred to as an *exothermic compound*. Benzene, which has an endothermic formation enthalpy, is called an *endothermic compound*.

How the exothermicity or endothermicity shows itself in practice depends on the reaction conditions. For instance, if the reaction is carried out in a thermally insulated container, then an endothermic reaction gets its necessary energy from the contents of the vessel; as a result, they cool. In contrast, an exothermic reaction in a thermally insulated vessel releases its energy into the contents of the vessel, and so their temperature rises. That is why water gets hot when concentrated sulphuric acid is added, because the dilution process is exothermic. If an endothermic reaction is run in a vessel in a water-bath that maintains constant temperature, then it draws in energy from the bath; conversely, an exothermic reaction releases energy into the bath.

Enthalpies of atomization

The *enthalpy of atomization*, $\Delta H_a^\ominus$, is the enthalpy change when a substance is shattered into a gas of atoms. There are two important cases. One is the *dissociation* of molecules. For example, the dissociation of chlorine is the atomization

$$\tfrac{1}{2}Cl_2(g) \rightarrow Cl(g); \qquad \Delta H_{a,m}^\ominus = +122 \text{ kJ mol}^{-1} \quad \text{at } 25\,°C.$$

The enthalpy of atomization of a diatomic molecule is half the enthalpy of dissociation of the molecule. The other important case is the *sublimation* of solid elements. For example, the sublimation of sodium metal is the atomization

$$Na(s) \rightarrow Na(g); \qquad \Delta H_{a,m}^\ominus = +108 \text{ kJ mol}^{-1} \quad \text{at } 25\,°C.$$

Lattice enthalpies and the Born-Haber cycle

The *lattice enthalpy* is a measure of the strength of an ionic crystal lattice and is the enthalpy change when the lattice forms from a gas of ions. In the case of sodium chloride, for instance, it is the enthalpy change in the process

$$Na^+(g) + Cl^-(g) \rightarrow NaCl(s); \qquad \Delta H_m^\ominus = -788 \text{ kJ mol}^{-1}.$$

This is a huge quantity of energy: it is enough to heat 2.5 dm^3 of water from room temperature to boiling.

The lattice enthalpy can be calculated from other data by using an adaptation of Hess's law known as the *Born-Haber cycle*. Consider the case of sodium chloride. We know (from calorimetric measurements) its enthalpy of formation (-411 kJ mol^{-1}). This formation reaction can be thought of as taking place in several steps, Fig. 9.10, one of which is the collapse of a gas of ions into the lattice. This complete circle of reactions

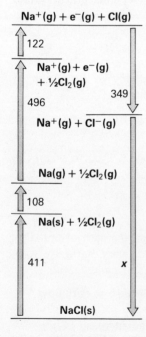

$$Na^+(g) + e^-(g) + Cl(g)$$

122

$$Na^+(g) + e^-(g) + \tfrac{1}{2}Cl_2(g)$$

496   349

$$Na^+(g) + Cl^-(g)$$

$$Na(g) + \tfrac{1}{2}Cl_2(g)$$

108

$$Na(s) + \tfrac{1}{2}Cl_2(g)$$

411   *x*

$$NaCl(s)$$

9.10 A Born-Haber cycle

is called a *cycle*. The first step is the atomization of sodium metal; the second is the ionization of its atoms; the third is the atomization of chlorine; the fourth is the attachment of the electron released by the sodium on to the chlorine atom; the fifth is the condensation of the ions into the lattice.

Example: Use thermochemical data to find the lattice enthalpy of sodium chloride.

Method: Construct a Born-Haber cycle which includes the NaCl formation reaction and the formation of the lattice from a gas of Na^+ and Cl^- ions. Use Table 1.2 for the ionization energy of Na, $+496 \, kJ \, mol^{-1}$, and the electron affinity of Cl, $+349 \, kJ \, mol^{-1}$. The atomization enthalpy of Na(s), $+108 \, kJ \, mol^{-1}$, and the atomization enthalpy of $Cl_2(g)$, $+122 \, kJ \, mol^{-1}$, are given in the text.

Answer: A suitable Born-Haber cycle is shown in Fig. 9.10. The unknown quantity is x, and on the basis that the distance up on the left must be the same as the distance down on the right, we have:

(distance up on left)/$kJ \, mol^{-1} = 411 + 108 + 496 + 122 = 1137$

(distance down on right)/$kJ \, mol^{-1} = 349 + x$

Therefore $x = 788$. It follows that the lattice enthalpy is $\Delta H_m^\ominus = -788 \, kJ \, mol^{-1}$, the sign indicating the downward direction of the last step (energy release).

Comment: The electron affinity of atoms is a difficult quantity to measure, therefore it is often taken as the unknown in a cycle of the kind treated here. This is how some of the values obtained in Table 1.2 were obtained. Others were obtained by calculation using quantum mechanics.

Bond enthalpies

Scientists are always on the alert for patterns emerging in the data they collect. Patterns emerge in thermochemistry, and when they are noticed they help in the applications of the data and in the understanding of the underlying molecular properties.

The enthalpy change in a reaction arises mainly from the changes that accompany the breaking of old bonds and the formation of new ones. Therefore, if we could compile a table of *bond enthalpies* we would be able to predict the enthalpy changes on the basis of the bonds that are changed.

The procedure runs into difficulty as soon as we start, as can be seen in the case of water. The overall enthalpy of formation is twice the enthalpy of atomization of H_2, plus the enthalpy of atomization of O_2, plus the enthalpy of formation of two O—H bonds, plus the enthalpy of condensation to the liquid. Now it is known (from spectroscopy) that different quantities of energy are needed to form O—H and to form H—OH:

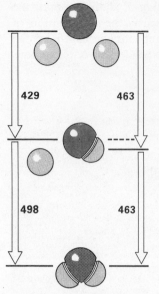

429   463

498   463

9.11 Bond enthalpies
and mean bond
enthalpy

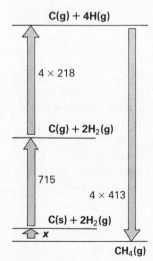

C(g) + 4H(g)

4 × 218

C(g) + 2H$_2$(g)

715

4 × 413

C(s) + 2H$_2$(g)

x

CH$_4$(g)

9.12 Estimating the
enthalpy of formation
of methane

spectroscopic measurements show that the enthalpy change of the first step is -429 kJ mol^{-1} but -498 kJ mol^{-1} for the second, Fig. 9.11. Therefore we cannot list an enthalpy of forming an O—H bond without specifying what other atoms are already present.

The way round this problem is to draw up a list of *mean bond enthalpies*, and to use them in *approximate* calculations of reaction enthalpies. The mean O—H bond enthalpy is -463 kJ mol^{-1} if the average is taken over a broad range of species. A list is given in Table 9.4. By adding together the appropriate values the enthalpies of reactions may be estimated.

Table 9.4   Mean bond enthalpies/kJ mol^{-1}

H—C	413	C—N	292
H—N	391	C—O	351
H—O	463	C—Cl	328
C—C	348	C=O	743
C=C	612		

Example: Estimate the molar enthalpy of formation of methane on the basis of mean bond enthalpies.

Method: The formation reaction is $C(s) + 2H_2(g) \rightarrow CH_4(g)$. In order to estimate the overall enthalpy change we have first to atomize carbon ($\Delta H^{\ominus}_{a,m} = +715$ kJ mol^{-1}) and dissociate $2H_2$ ($\Delta H^{\ominus}_{a,m} = +218$ kJ mol^{-1}), and then form 4 C—H bonds (mean bond enthalpies in Table 9.4). The overall process is represented by the cycle shown in Fig. 9.12.

Answer: The (distance up on the left)/kJ mol^{-1} = $x + 715 + 4(218) = x + 1587$. The (distance down on the right)/kJ mol^{-1} = $4(413) = 1652$. Hence $x = (1652 - 1587) = 65$. Therefore the molar enthalpy of formation of methane, the negative of x in the diagram, is -65 kJ mol^{-1}.

Comment: The experimental value is -74.8 kJ mol^{-1}. The use of mean bond enthalpies is an approximation, but it is a useful and easy way of estimating data that may not be to hand.

Delocalization enthalpy

When thermochemical arguments are applied to benzene (and to other aromatic molecules) strange results are obtained. At least, they were strange at the time: our familiarity with the stability of aromatic rings has removed the striking quality of the results obtained by the early chemists.

The enthalpy of formation of benzene can be predicted by adding together the appropriate mean bond enthalpies; this is done in Fig. 9.13 on the basis that it is a ring of alternating double and single bonds. The answer is $+209$ kJ mol^{-1}. The experimental value is $+49$ kJ mol^{-1}, indicating that the calculation has *underestimated* its thermochemical stability by 160 kJ mol^{-1}, a chemically very significant value.

We now know, and we saw in Chapter 2, that it is too naive to regard a benzene molecule as a ring of alternating single and double bonds; in fact the molecule is stabilized by the delocalization of the electrons. Thermochemistry provides a quantitative measure of this *delocalization stabilization* and has shown that the delocalization reduces the enthalpy of the molecule by as much as 160 kJ mol^{-1}. This quantity is called the *delocalization enthalpy* of benzene. A similar discrepancy between observation and expectation is provided by measurements of the *enthalpies of hydrogenation* of benzene and cyclohexene, as demonstrated in the following *Example*.

Example: The enthalpy of hydrogenation of cyclohexene is -120 kJ mol^{-1}; the enthalpy of hydrogenation of benzene is -208 kJ mol^{-1}. Calculate the delocalization enthalpy of benzene.

Method: The two reactions are

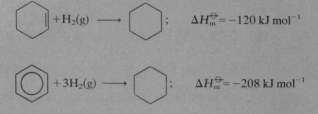

Calculate the hydrogenation enthalpy of benzene on the basis that it has three localized double bonds. The difference between this value and the observed value is the delocalization enthalpy, Fig. 9.13.

Answer: The calculated hydrogenation enthalpy is $3(-120$ kJ mol$^{-1}) = -360$ kJ mol^{-1}. The delocalization enthalpy is therefore the difference $\Delta H_{\text{deloc}} = (-208$ kJ mol$^{-1}) - (-360$ kJ mol$^{-1}) = 152$ kJ mol^{-1}.

Comment: Since a smaller enthalpy change occurs when benzene is hydrogenated than would be expected on the basis that it has three double bonds, we conclude that it is more stable than the latter structure suggests. The value 152 kJ mol^{-1} represents the thermochemical stabilization brought about by delocalization of the bonds. Note that slightly different values of ΔH_{deloc} are obtained depending on the basis of the calculation: this shows that the delocalization enthalpy is not a well defined characteristic of the molecule.

6C(g) + 6H(g)

1308
1044
form 3 C—C bonds

6C(g) + 3H$_2$(g)

4290
1836
form 3 C=C bonds

2478

form 6 C—H bonds

6C(s) +3H$_2$(g)

160

Benzene (g)

Benzene (l) 31

6C(s)+3H$_2$(g) 49

9.13 The delocalization enthalpy of benzene

Summary

1 *Thermodynamics* is the study of the transformations of energy; *thermochemistry* is the branch of thermodynamics concerned with transformations of energy in the course of chemical reactions.

2 The *First law of thermodynamics* states that energy can be neither created nor destroyed.

3 Energy can be transferred from one system to another as *heat* or as *work*.

4 If a quantity of energy, *w*, is transferred *to* a system as work and a quantity of energy, *q*, is transferred *to* the system as heat, the transfer changes the *internal energy, U*, by $\Delta U = q + w$.

5 When a quantity of energy, *q*, is transferred to a *constant volume system* as heat, the internal energy changes by $\Delta U = q$.

6 The *enthalpy, H*, of a system is a modification of the internal energy which automatically takes into account the work of expansion when a system is heated at constant pressure.

7 When a quantity of energy, *q*, is transferred to a *constant pressure system* as heat, the enthalpy changes by $\Delta H = q$.

8 The *molar enthalpy of vaporization* is the change of enthalpy that occurs when unit amount of substance of a liquid evaporates; it is the energy that has to be supplied as heat when vaporization occurs at constant pressure.

9 The *molar enthalpy of melting* (or fusion) is the change of enthalpy that occurs when unit amount of substance of a solid melts; it is the energy that has to be supplied as heat when the melting occurs at constant pressure.

10 Enthalpies of vaporization and melting are positive; the *enthalpy of freezing* is the negative of the enthalpy of melting.

11 The experimental method of measuring transfers of energy as heat is called *calorimetry*. The *flame calorimeter* operates at constant pressure, and so gives the enthalpy of reaction directly (via the temperature change and the heat capacity of the calorimeter). The *bomb calorimeter* operates at constant volume, and so gives the change of internal energy during the reaction.

12 The *enthalpy of solution* is the enthalpy change accompanying dissolution of a salt to give an infinitely dilute solution; the *enthalpy of neutralization* is the enthalpy change accompanying neutralization of an acid by a base in infinitely dilute solution.

13 The *enthalpy of combustion* is the enthalpy change accompanying the complete combustion of a sample (to carbon dioxide and water in the case of organic materials such as hydrocarbons and carbohydrates).

14 The *standard state* of a substance is its most stable form under a pressure of 1 atm (101.325 kPa) at the temperature specified.

15 The *standard enthalpy of reaction*, $\Delta H^{\ominus}$, is the enthalpy of reaction when all the reactants and all the products are in their standard states.

16 The standard molar *enthalpy of formation*, $\Delta H_{f,m}^{\ominus}$, is the enthalpy change accompanying formation of unit amount of the compound from its elements, all components being in their standard states.

17 The standard *enthalpy of atomization*, $\Delta H_a^{\ominus}$, is the enthalpy change accompanying the decomposition of a substance into a gas of atoms.

18 *Hess's law* states that the overall reaction enthalpy is the sum of the reaction enthalpies of each step into which the reaction may formally be divided.

19 A *Born-Haber cycle* is a complete sequence of reactions involving a lattice enthalpy from an initial set of elements and back to them through a different route.

20 If $\Delta H < 0$, a reaction is *exothermic*; if $\Delta H > 0$, a reaction is *endothermic*.

21 A *compound* is endothermic if its enthalpy of formation is positive, but exothermic if it is negative.

22 The *mean A—B bond enthalpy* is the average of the enthalpy changes accompanying the breaking of the A—B bond in a series of compounds.

23 The *thermochemical stabilization* of aromatic compounds is indicated by their *delocalization enthalpies*. The delocalization enthalpy of benzene is the difference between its experimental enthalpy of formation and that calculated from mean bond enthalpies on the basis that it is built from a ring of alternating double and single bonds.

Problems

1 Explain the terms *work, heat,* and *internal energy,* and state the *First law of thermodynamics.*

2 Explain why enthalpy changes are so important in chemistry.

3 The standard molar enthalpy of vaporization of benzene at its boiling temperature is 30.8 kJ mol^{-1}: for how long would a 100 W electric heater have to operate in order to vaporize a 100 g sample at that temperature?

4 Suppose you were caught on a mountain with soaking wet clothes; what energy would your body have to supply if the wind caused the evaporation of 500 g of water?

5 Foodstuffs are often classified as 'so many Calories': given that 1 Calorie (as used by dieticians) corresponds to 4.18 kJ, calculate (to one significant figure) the mass of cheese that must be consumed in order to offset the energy loss in the last question, given that the calorific value of cheese is 5.0 Calories per gram.

6 The calorific value of milk is about 2.9 kJ g^{-1}. To what height could a person (mass 70 kg) climb above the surface of the earth in the course of using the entire energy one pint can supply?

7 Explain the terms *standard state* of a substance and *standard molar enthalpy of formation.*

8 Explain the connection between Hess's law and the First law of thermodynamics. What is the evidence for both laws?

9 Use the standard molar enthalpies of formation in Table 9.3 to calculate the standard enthalpy changes accompanying the following reactions: (*a*) $CH_4(g) + 2O_2(g) \rightarrow CO_2(g) + 2H_2O(l)$, (*b*) $4Al(s) + 3O_2(g) \rightarrow 2Al_2O_3(s)$.

10 Calculate the enthalpy of dissolution of sodium chloride (the enthalpy change for $NaCl(s) \rightarrow NaCl(aq)$) on the basis that the enthalpies of solvation of $Na^+(g)$ and $Cl^-(g)$ are -406 kJ mol^{-1} and -364 kJ mol^{-1}, respectively. (Recall the discussion in Chapter 8.)

11 Present a thermochemical argument to explain why sodium chloride does not dissolve in benzene.

12 (*a*) State Hess's law of constant heat summation (law of conservation of energy).

   (*b*) Define precisely the terms
      (*i*) *enthalpy of formation* (*heat of formation*);

(*ii*) *enthalpy of neutralization* (*heat of neutralization*).

(*c*) Using the following data,[1] collected at 25 °C and standard atmospheric pressure, in which the negative sign indicates heat evolved, calculate the enthalpy of formation of potassium chloride, KCl(s):

	ΔH(298 K)/kJ mol^{-1}
(*i*) $KOH(aq) + HCl(aq) = KCl(aq) + H_2O(l)$	-57.3
(*ii*) $H_2(g) + \frac{1}{2}O_2(g) = H_2O(l)$	-285.9
(*iii*) $\frac{1}{2}H_2(g) + \frac{1}{2}Cl_2(g) + aq = HCl(aq)$	-164.2
(*iv*) $K(s) + \frac{1}{2}O_2(g) + \frac{1}{2}H_2(g) + aq = KOH(aq)$	-487.0
(*v*) $KCl(s) + aq = KCl(aq)$	$+18.4$

[1] Candidates from certain schools may have used (c) instead of (s) to denote the solid (crystalline) state.

(*d*) The enthalpy of neutralization of ethanoic acid (acetic acid) is -55.8 kJ mol^{-1} while that of hydrochloric acid is -57.3 kJ mol^{-1}, both reactions being with potassium hydroxide solution. Explain the difference in these two values and make what deductions you can.

(*e*) Describe briefly an experiment by which the enthalpy change of a chemical reaction *of your own choice* may be determined, and outline how the various measurements would be used to calculate the final value.

(*f*) When solid potassium chloride is dissolved in water, heat is absorbed.

$$KCl(s) + aq = KCl(aq); \quad \Delta H = +18.4 \text{ kJ mol}^{-1}$$

Use the ideas formulated in the kinetic-molecular theory of matter to discuss what happens during the dissolution of a crystalline salt in water, and explain why heat is absorbed in the above reaction.

(*SUJB*)

13 (*a*) Use the following information to calculate (*i*) the enthalpy of formation of methane from graphite and hydrogen in their standard states, (*ii*) the C—H bond energy [enthalpy] in the methane molecule. (All data refer to 298 K.)

$CH_4(g) + 2O_2(g) = CO_2(g) + 2H_2O(l)$	$\Delta H = -890 \text{ kJ mol}^{-1}$
$C(graphite) + O_2(g) = CO_2(g)$	$\Delta H = -394 \text{ kJ mol}^{-1}$
$H_2(g) + \frac{1}{2}O_2(g) = H_2O(l)$	$\Delta H = -286 \text{ kJ mol}^{-1}$
$C(graphite) = C(g)$	$\Delta H = \quad 717 \text{ kJ mol}^{-1}$
$H_2(g) = 2H(g)$	$\Delta H = \quad 436 \text{ kJ mol}^{-1}$

(*b*) State what information you would need about ethane such that, together with the above data, you could make an estimate of the C—C bond energy in the ethane molecule. (An explanation of your answer is not required.)

(*Oxford*)

14 (*a*) State the First law of thermodynamics.

(*b*) The standard enthalpies of formation, $\Delta H_f^{\ominus}$, for methane, carbon monoxide, carbon dioxide and water are in the table below.

	$\Delta H_f^{\ominus}$ in kJ mol^{-1}
Methane	-74.9
Carbon monoxide	-111
Carbon dioxide	-394
Water	-286

Principles of physical chemistry

The molar volume of a gas is $22.4 \, \text{dm}^3$ $(2.24 \times 10^{-2} \, \text{m}^3) \, \text{mol}^{-1}$ at s.t.p. Calculate the heat evolved when $1 \, \text{m}^3$ of methane, measured at s.t.p., is burnt completely.

Compare the value obtained with the value for burning completely $1 \, \text{m}^3$, measured at s.t.p., of coal gas, the composition by volume of which is 50% hydrogen, 35% methane and 15% carbon monoxide.

What advantage does natural gas, which contains about 90% methane, have over coal gas?

(c) The following data refer to the reaction of aqueous hydrochloric acid with sodium hydroxide.

$$HCl(aq) + NaOH(aq) \rightarrow NaCl(aq) + H_2O(l) \qquad \Delta H^{\ominus} = -57.1 \, \text{kJ mol}^{-1}$$
$$HCl(aq) + NaOH(s) \rightarrow NaCl(aq) + H_2O(l) \qquad \Delta H^{\ominus} = -99.8 \, \text{kJ mol}^{-1}$$

Account qualitatively for the differences in the values of the enthalpy changes given by referring to the processes that occur when solid sodium hydroxide dissolves in water.

(AEB 1979)

15 State the First law of thermodynamics in terms of *heat* and *work*, defining these quantities carefully.

Use the bond enthalpies [in Table 9.4, plus the values of $366 \, \text{kJ mol}^{-1}$ and $276 \, \text{kJ mol}^{-1}$ for H-Br and C-Br, respectively] to calculate the standard enthalpy change for the reaction

$$C_2H_4(g) + HBr(g) \rightarrow C_2H_5Br(g).$$

Given that the molar enthalpy of vaporization of bromoethane is $27.0 \, \text{kJ mol}^{-1}$, compare your calculated value of $\Delta H^{\ominus}(298 \, \text{K})$ with the one calculated from the standard enthalpies of formation [$-85.4 \, \text{kJ mol}^{-1}$ for $C_2H_5Br(l)$] and comment on your answer.

(Oxford and Cambridge S part question)

16 Two possible ways of fixing atmospheric nitrogen may be represented by the equations:

$$\tfrac{1}{2}N_2(g) + \tfrac{3}{2}H_2O(g) \rightarrow NH_3(g) + \tfrac{3}{4}O_2(g)$$
$$\tfrac{1}{2}N_2(g) + \tfrac{1}{2}H_2O(g) + \tfrac{5}{4}O_2(g) \rightarrow HNO_3(l).$$

Calculate the value of $\Delta H^{\ominus}$ for each of these processes at 298 K.

Why do you think that the process represented by the second equation does not provide a suitable method of fixing nitrogen?

(Oxford and Cambridge S part question)

17 (a) (i) Define enthalpy of formation ΔH_f of a compound.
　　　　(ii) What extra conditions must be imposed to specify the standard enthalpy of formation $\Delta H_f^{\ominus}$ of a compound?
　　(b) When ethanol burns in oxygen, carbon dioxide and water are formed.
　　　　(i) Write the equation which describes this reaction.
　　　　(ii) Using the data

$$\Delta H_f^{\ominus} \text{ for ethanol (l)} \qquad = -277.0 \, \text{kJ mol}^{-1}$$
$$\Delta H_f^{\ominus} \text{ for carbon dioxide (g)} = -393.7 \, \text{kJ mol}^{-1}$$
$$\Delta H_f^{\ominus} \text{ for water (l)} \qquad = -285.9 \, \text{kJ mol}^{-1}$$

calculate the value of $\Delta H^{\ominus}$ for the combustion of ethanol. *(JMB)*

18 What factors control the magnitude of lattice energies [enthalpies]? What information can be obtained from a comparison of calculated lattice energies with those obtained from a Born-Haber cycle?

Given the following quantities (in kJ/mol) for rubidium chloride, obtain a value for the electron affinity of the Cl atom:

Lattice energy of RbCl	665
Dissociation energy of Cl_2 (gas)	226
Heat [enthalpy] of sublimation of Rb metal	84
Ionization energy of the Rb atom	397
Standard heat [enthalpy] of formation of solid RbCl	−439

(*Cambridge Entrance*)

10 Reactions at equilibrium

One of the most important pieces of information in chemistry is whether a reaction will run in one direction or another. We first describe what it means to say that a chemical reaction is at equilibrium, and see that the composition at equilibrium can be specified in terms of a single quantity, the equilibrium constant. We shall also see how it is possible to predict how the position of equilibrium responds to changes in the conditions, and go on to show how it is also possible to make quantitative predictions.

Introduction

Some reactions seem to go just so far and then stop, even though there are plenty of reactants left. Others seem to go all the way to products. An example of the first type is the synthesis of ammonia; a mixture of nitrogen and hydrogen reacts under suitable conditions and forms ammonia, but the production of ammonia comes to a halt at some stage even though there is plenty of unreacted hydrogen and nitrogen. As an example of the second type, take the closely related reaction of hydrogen with oxygen to give water. When they are induced to react (by a spark, for instance), they go completely to the product.

These examples raise several questions. In the first place, are there really two sorts of reaction, one going to completion and the other not? In the second place, how should we describe the position of chemical equilibrium? We shall see that this can be done through the *equilibrium constant* of the reaction. Once we have the values of the equilibrium constant we can discuss the equilibrium compositions of reaction mixtures and their response to the conditions, such as the temperature and the pressure. This point of view is developed in the following three chapters, and we shall see that the same style of argument can be applied to a wide range of phenomena.

10.1 The description of equilibrium

When a reaction reaches *equilibrium*, when the composition is steady and has no further tendency to react, it is not just dead. Take, for example, an ammonia synthesis that has settled down to a stationary composition consisting of a mixture of nitrogen, hydrogen, and ammonia. The overall composition of the reaction mixture does not change, but individual molecules go on reacting in such a way that the average rate of formation of products is exactly balanced by their decay. In other words, chemical equilibrium is *dynamic* equilibrium. This can be proved by isotopic studies: if some N^2H_3 is added to the equilibrium mixture, in due course deuterium (^2H) appears in the molecular hydrogen as $^2H^1H$, and species such as $N^2H^1H_2$ appear in the ammonia.

When we turn to a reaction that 'goes to completion' we find exactly the same type of behaviour. Such reactions also reach an equilibrium, a stationary but dynamic condition in which the forward and backward reaction rates balance, but for them the equilibrium composition lies strongly in favour of the products. Even in a reaction that apparently 'goes to completion' there is always a tiny amount of reactant in the equilibrium mixture.

All reactions are the same in kind but different in degree; all reach dynamic equilibrium with both reactants and products present, but in some the equilibrium lies so strongly in favour of the products that we fail to notice the presence of the minute proportion of reactants.

The equilibrium constant

Many experiments to measure the compositions of reaction mixtures at equilibrium have been carried out. It has been discovered that the equilibrium composition can be characterized by a constant, K. The equilibrium constant can be measured experimentally, and its value can be predicted on the basis of thermodynamics (as we see in Chapter 11). Once its value is known for a reaction, it is possible to do a wide variety of useful calculations.

We shall introduce the equilibrium constant by considering a special case, the esterification of ethanoic acid, Fig. 10.1:

$$CH_3COOH(l) + C_2H_5OH(l) \rightleftharpoons CH_3COOC_2H_5(l) + H_2O(l).$$

This is an example of a *homogeneous* equilibrium, since all the compounds are in the same phase (liquid). From the data in Table 10.1 it can be seen how the equilibrium composition of the reaction mixture depends on the relative amounts of acid and alcohol mixed initially. Although the information may seem to have no pattern, one does emerge on looking more closely. The fifth column in Table 10.1 gives the value of the quantity

$$K_c = \left\{ \frac{[CH_3COOC_2H_5][H_2O]}{[CH_3COOH][C_2H_5OH]} \right\}_{eq} \tag{10.1.1}$$

where the [. . .] are the concentrations of the species and 'eq' denotes equilibrium; the c on K_c is used when the constant is expressed in terms of concentrations. K_c has about the same value, 4.0, whatever the initial composition of the mixture. It therefore characterizes the equilibrium position of this reaction (at $100\,^\circ C$) and in order to specify the equilib-

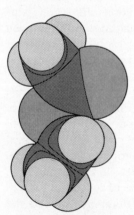

10.1 A molecular model of the esterification of ethanoic acid

Principles of physical chemistry

rium position of the reaction at 100 °C all we need do is quote the value of K_c.

Table 10.1 The esterification of ethanoic acid at 100 °C

Amounts of substance				
n_{acid}/mol at start	$n_{alcohol}$/mol at start	n_{ester}/mol at equilibrium	n_{water}/mol at equilibrium	K_c
1.00	0.18	0.171	0.171	3.9
1.00	0.33	0.293	0.293	3.3
1.00	0.50	0.414	0.414	3.4
1.00	1.00	0.667	0.667	4.0
1.00	2.00	0.858	0.858	4.5
1.00	8.00	0.966	0.966	3.9

Example: The analysis of the product composition of the equilibrium mixture of the esterification reaction showed that it contained 0.67 mol CH_3COOH, 0.82 mol C_2H_5OH, and 0.73 mol H_2O. What amount of substance of the ester $CH_3COOC_2H_5$ is present?

Method: At equilibrium the concentrations of the four species are related by eqn (10.1.1) with $K_c = 4.0$ at 100 °C. The volumes in the concentration terms all cancel, and so the amount of ester, n, may be obtained by using the expression in the form

$$n = \left\{ \frac{n(CH_3COOH) \times n(C_2H_5OH)}{n(H_2O)} \right\} \times K_c.$$

Answer: Inserting the data into the expression just quoted gives

$$n = \left\{ \frac{(0.67 \text{ mol}) \times (0.82 \text{ mol})}{(0.73 \text{ mol})} \right\} \times 4.0 = 3.0 \text{ mol}.$$

Comment: If the reaction had been run in a large quantity of water so that the equilibrium composition corresponded to 0.67 mol CH_3COOH, 0.82 mol C_2H_5OH in 1 dm³ water (55.5 mol H_2O), the same calculation leads to $n = 0.040$ mol, a huge decrease in ester (as a result of hydrolysis). The equilibrium constant is the same, but the individual compositions have adjusted. The equilibrium constant for this reaction barely changes with temperature in the region of 100 °C. We develop this *Example* later.

The general definition of the equilibrium constant, valid for any reaction, can be introduced in two steps. First, consider a homogeneous reaction where all the species are in solution (the esterification reaction is an example). Then it has the form

$$aA + bB \rightleftharpoons cC + dD; \qquad K_c = \left\{ \frac{[C]^c[D]^d}{[A]^a[B]^b} \right\}_{eq}. \qquad (10.1.2)$$

In the case of the esterification reaction, $a = b = c = d = 1$, and so we get eqn (10.1.1) again. We shall see some more examples later.

Now consider a homogeneous reaction in the *gas phase* (such as the synthesis of ammonia). Although we could express the equilibrium constant in terms of the concentrations of the gases present at equilibrium (using the last equation), it is more natural to express it in terms of the equilibrium partial pressures of the gases (and to denote it K_p). Therefore we write

$$aA(g) + bB(g) \rightleftharpoons cC(g) + dD(g); \quad K_p = \left\{ \frac{p_C^c p_D^d}{p_A^a p_B^b} \right\}_{eq}. \tag{10.1.3}$$

The p's are the partial pressures of the components.

As an example, consider the ammonia synthesis:

$$N_2(g) + 3H_2(g) \rightleftharpoons 2NH_3(g); \quad K_p = \{ p_{NH_3}^2 / p_{N_2} p_{H_2}^3 \}_{eq}, \tag{10.1.4}$$

since $a = 1$, $b = 3$, $c = 2$, and $d = 0$. At 400 K the value of this equilibrium constant is 53 atm^{-2}, and so we can begin to make predictions about the composition of the reaction mixture at equilibrium, and design the synthesis plant accordingly.

Using the equilibrium constant

The equilibrium constant specifies the equilibrium composition of reactions, but because the information is in the form of a *ratio* of products of concentrations or pressures, it is not always immediately obvious what that information is! We therefore illustrate how to extract the information by doing a couple of examples.

If in the esterification reaction at equilibrium the amount of ester present is n, then the initial amounts of acid and alcohol must each have been reduced by an amount n. This can be set out clearly by drawing up the following table:

	Acid CH$_3$COOH	Alcohol C$_2$H$_5$OH	Ester CH$_3$COOC$_2$H$_5$	Water H$_2$O
Initial amounts of substance	n_{ac}	n_{al}	0	0
Equilibrium amounts of substance	$n_{ac} - n$	$n_{al} - n$	n	n
Concentration at equilibrium	$\dfrac{n_{ac} - n}{V}$	$\dfrac{n_{al} - n}{V}$	$\left(\dfrac{n}{V} \right)$	$\left(\dfrac{n}{V} \right)$

By substituting the equilibrium concentrations into the expression for the equilibrium constant, eqn (10.1.1), we get

$$K_c = \left\{ \frac{(n/V) \cdot (n/V)}{[(n_{ac} - n)/V][(n_{al} - n)/V]} \right\} = \frac{n^2}{(n_{ac} - n)(n_{al} - n)}. \tag{10.1.5}$$

This can be solved for n (it can be turned into a quadratic equation), and so the equilibrium composition can be predicted simply by feeding in the appropriate values of the initial composition and the equilibrium constant.

Example: $50 \, \text{cm}^3$ of ethanoic acid is mixed with $50 \, \text{cm}^3$ of ethanol. What is the composition of the mixture at $100 \, °C$?

Method: Use eqn (10.1.5). Find n_{ac} and n_{al} by converting the volumes to masses, using the densities $1.049 \, \text{g cm}^{-3}$ (acid) and $0.789 \, \text{g cm}^{-3}$ (alcohol), and then to amounts of substance using $M_r = 60.05$ (acid) and 46.07 (alcohol). Rearrange eqn (10.1.5) into a quadratic equation in n, and solve using the standard formula for quadratic equations of the form $an^2 + bn + c = 0$; that is, $n = (1/2a)\{-b \pm \sqrt{(b^2 - 4ac)}\}$ and select the correct root. Form the amounts of substance by calculating $n_{ac} - n$, $n_{al} - n$, n, n for the amounts of acid, alcohol, ester, and water present at equilibrium. Use $K_c = 4.0$.

Answer: $n_{ac} = \dfrac{(50 \, \text{cm}^3) \times (1.049 \, \text{g cm}^{-3})}{(60.05 \, \text{g mol}^{-1})} = 0.873 \, \text{mol}$

$n_{al} = \dfrac{(50 \, \text{cm}^3) \times (0.789 \, \text{g cm}^{-3})}{(46.07 \, \text{g mol}^{-1})} = 0.856 \, \text{mol}.$

Equation (10.1.5) rearranges into

$$n^2 - (n_{ac} - n)(n_{al} - n)K_c = 0, \quad \text{or} \quad (1 - K_c)n^2 + (n_{ac} + n_{al})K_c n - n_{ac}n_{al}K_c = 0.$$

Therefore, for an equation of the form $an^2 + bn + c = 0$

$a = 1 - K_c = 1 - 4.0 = -3.0$

$b = (n_{ac} + n_{al})K_c = (0.873 \, \text{mol} + 0.856 \, \text{mol}) \times 4.0 = 6.92 \, \text{mol}$

$c = -n_{ac}n_{al}K_c = -(0.873 \, \text{mol}) \times (0.856 \, \text{mol}) \times 4.0 = -2.99 \, \text{mol}^2$

$n = (1/2a)\{-b \pm \sqrt{(b^2 - 4ac)}\}$

$= \dfrac{(-6.92 \, \text{mol}) \pm \sqrt{\{(6.92 \, \text{mol})^2 - 4 \times (-3.0) \times (-2.99 \, \text{mol}^2)\}}}{2 \times (-3.0)}$

$= \dfrac{(-6.92 \, \text{mol}) \pm (3.47 \, \text{mol})}{(-6.0)} = 0.58 \, \text{mol or } 1.73 \, \text{mol}.$

The second answer is impossible (because no more than $0.856 \, \text{mol}$ of product can be generated with the specified starting amounts), and so we conclude that at equilibrium $n = 0.58 \, \text{mol}$. It follows that the equilibrium composition is $0.58 \, \text{mol} \, CH_3COOC_2H_5$, $0.29 \, \text{mol} \, CH_3COOH$, $0.28 \, \text{mol} \, C_2H_5OH$, and $0.58 \, \text{mol} \, H_2O$.

Comment: An interesting exercise is to suppose that water is also present initially (*i.e.*, the reaction was run in water). Rework the calculation on the assumption that 1 kg water was also present initially.

As a second illustration, consider the description of the gas-phase equilibrium between N_2O_4 and $2NO_2$. Above $22 \, °C$ the former is a colourless gas, while the latter is brown. Since one species is coloured and the other is not, it is easy to follow the composition of the vapour spectroscopically.

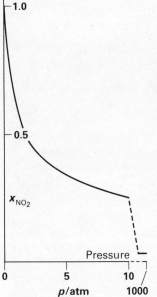

10.2 The mole fraction of NO_2 in the $N_2O_4 \rightleftharpoons 2NO_2$ equilibrium at 25 °C

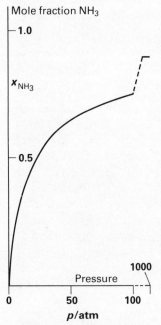

10.3 The mole fraction of NH_3 in the $N_2 + 3H_2 \rightleftharpoons 2NH_3$ equilibrium at 400 °C

The reaction equilibrium is

$$N_2O_4(g) \rightleftharpoons 2NO_2(g); \qquad K_p = \{p^2_{NO_2}/p_{N_2O_4}\}_{eq}. \qquad (10.1.6)$$

It follows from the equation that for every N_2O_4 molecule falling apart, two NO_2 molecules are formed. The same kind of calculation used for the esterification equilibrium leads to the result that the mole fraction of NO_2 depends on the total pressure as shown in Fig. 10.2. This shows that as the total pressure is increased, the mole fraction of nitrogen dioxide decreases, and at very high pressures there is very little present at all. It is as though the NO_2 molecules have been squashed back into dimers.

The similar but contrasting cases of the syntheses of ammonia and water can be dealt with in the same way. If nitrogen and hydrogen are present in the mole ratio 1:3 and this mixture is allowed to come to equilibrium, then the mole fraction of ammonia depends on the total pressure of the mixture as shown in Fig. 10.3. In contrast to the case of $N_2O_4 \rightleftharpoons 2NO_2$, an increase in pressure drives the equilibrium to the right. A pattern is beginning to emerge.

When the technique is applied to the water synthesis reaction under conditions where all the water produced is a vapour, its mole fraction in the equilibrium mixture depends on the total pressure as shown in Fig. 10.4.

These results show that the differences between the equilibrium compositions of the ammonia and water synthesis reactions are of degree rather than of kind. First, the pressure dependence of the equilibrium is very similar in the two cases; the curves drawn in Figs 10.3 and 10.4 differ in detail, but have about the same shape. Quantitatively, though, at any given pressure the equilibrium compositions are very different. When the total pressure is 1 atm, the mole fraction of ammonia at equilibrium is 0.17, a readily measurable but not a dominating quantity. In contrast, under the same conditions, the mole fraction of water is so close to 1 that, for all practical purposes, only water vapour is present. In other words, the synthesis equilibrium lies overwhelmingly in favour of the products in the case of water, but slightly in favour of the reactants in the case of ammonia. Qualitatively they are very similar; quantitatively they are very different.

Equilibria involving solids

When a chemical reaction involves species in different phases, such as a solid decomposing into another solid plus a gas, the stationary composition is called a *heterogeneous equilibrium*. An example is the decomposition of calcium carbonate into calcium oxide plus carbon dioxide:

$$CaCO_3(s) \rightleftharpoons CaO(s) + CO_2(g); \qquad K = \{[CaO]p_{CO_2}/[CaCO_3]\}_{eq}.$$

The equilibrium constant is expressed in terms of the concentrations of the solids and the pressure of the gas. But what is meant by the concentration of a pure solid? We express concentration as *amount of substance per unit volume* (which in practice means mol dm^{-3}), but for a pure solid amount of substance per unit volume is proportional to mass per unit volume, or *density*. The density of a solid is independent of how much of it is present, therefore the 'concentration' of a pure solid is also a constant. It is obviously sensible to combine any constants with K itself, and so from now on we shall write

$$CaCO_3(s) \rightleftharpoons CaO(s) + CO_2(g); \qquad K_p = \{p_{CO_2}\}_{eq}. \qquad (10.1.7)$$

Similar arguments apply to any heterogeneous equilibrium involving solids.

At 800 °C the equilibrium constant has the value $K_p = 24.4$ kPa. Therefore when calcium carbonate is heated to 800 °C in a closed vessel and allowed to come to equilibrium the pressure of carbon dioxide is 24.4 kPa. In this case K_p is simply the *dissociation pressure* of calcium carbonate at 800 °C. This shows that we can easily measure the equilibrium constant by noting the pressure exerted by a sample heated to the temperature of interest. When calcium carbonate is heated to 800 °C in an open vessel it will continue to decompose until the partial pressure of carbon dioxide reaches its equilibrium pressure, 24.4 kPa. On the other hand, since the gas is free to escape, the local partial pressure will never reach that value, because the carbon dioxide simply blows away into the surroundings. Therefore the carbonate will continue to decompose, and the reaction will stop when only calcium oxide remains.

This should seem like common sense. The only difference from conventional common sense is that we have been able to express the qualitatively obvious in *quantitative* terms, because a knowledge of the equilibrium constant lets us predict the magnitudes of the pressures involved. This is a principal feature of science: it is basically common sense equipped with numbers.

10.2 The effect of conditions

In this section we look at some of the ways of controlling the position of equilibria. This is enormously important when thinking about the design of an industrial plant or how a biological cell contributes to the activity of an organism. For instance, what is the pressure of oxygen in equilibrium with blood cells inside the lung? Is it better to run the ammonia synthesis reaction at a high temperature? Since the Haber-Bosch synthesis of ammonia (which emerged under the pressures of war and the demand for explosives) is now the basis of the fertilizer industry, there is obviously great human and economic impact to gain from maximizing the yield.

Le Châtelier's principle

A simple rule lets us predict what happens to the position of chemical equilibrium when the conditions are changed. It was formulated by le Châtelier (in 1888) on the basis of experimental observations.

> *Le Châtelier's principle: When a reaction at equilibrium is subjected to a change of conditions, the composition adjusts so as to minimize the change.*

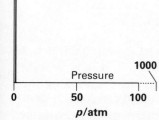

10.4 The mole fraction of H_2O in the $O_2 + 2H_2 \rightleftharpoons 2H_2O$ equilibrium at 400 °C

We shall apply the principle to the three basic types of change: changes of concentration, of pressure, and of temperature. We shall first show how the principle lets us make *qualitative* predictions, and then go on to justify the principle in terms of the equilibrium constant. This second step also allows us to make *quantitative* predictions.

How equilibria respond to changes of concentration

The central point to remember is that *the value of the equilibrium constant is independent of the concentrations.* This means that the composition will always adjust so as to keep the value of K_c constant.

Consider the esterification and hydrolysis equilibrium

$$CH_3COOH + C_2H_5OH \rightleftharpoons CH_3COOC_2H_5 + H_2O.$$

Suppose we pour some water into the equilibrium mixture. According to le Châtelier's principle, the equilibrium shifts so as to minimize the increase in the water concentration. Therefore we predict that the equilibrium will shift to the left, and the ester will be hydrolysed.

The qualitative result is made quantitative using the expression for the equilibrium constant,

$$K_c = \left\{ \frac{[CH_3COOC_2H_5][H_2O]}{[CH_3COOH][C_2H_5OH]} \right\}_{eq} = 4.0 \quad \text{at } 100\,°C.$$

In order to maintain the ratio of concentrations constant when the concentration of water is artificially increased (so that the numerator increases), the ester concentration must fall, and the denominator must increase. In other words, the equilibrium shifts back to favour the reactants, in accord with le Châtelier's principle. Since the ratio of products must continue to equal 4.0, we can also state *quantitatively* the effect of the addition of a known amount of water.

How equilibria respond to pressure

We shall limit the discussion to reactions involving gases, because only they respond significantly to changes of pressure (unless we deal with very high pressures, as under geological conditions). According to le Châtelier's principle, a reaction at equilibrium responds to an increase of pressure by minimizing the increase.

Take as an example the synthesis of ammonia,

$$N_2(g) + 3H_2(g) \rightleftharpoons 2NH_3(g); \qquad K_p = \{p_{NH_3}^2 / p_{N_2} p_{H_2}^3\}_{eq}.$$

Two molecules of ammonia are produced by consuming four molecules of gaseous reactants (one N_2 and three H_2). If the reaction is in a constant volume vessel, the reduction of the number of particles caused by the forward reaction results in a decrease in the pressure. Therefore, according to le Châtelier's principle, the equilibrium adjusts to an increase of applied pressure by shifting towards ammonia.

Figure 10.3 expresses this conclusion quantitatively. It shows that the gas becomes richer in ammonia as the pressure is increased. Therefore, the yield from an ammonia synthesis plant can be improved by operating it at high pressures. The reverse applies to the N_2O_4 dissociation, for the forward reaction increases the number of particles, and therefore corresponds to an increase in pressure. Le Châtelier's principle therefore predicts that an increase of pressure favours a shift of the equilibrium position to the left, in favour of the dimer.

A general reaction can now be treated very easily. If the number of gas-phase molecules on the right of the equation is smaller than on the left, then an increase of pressure results in the equilibrium shifting to the right, because in that way the system tends to minimize the increase of pressure. If there are fewer molecules on the left an increase in pressure shifts the equilibrium to the left. This can be summarized as follows:

When the pressure is increased, an equilibrium involving gases shifts as shown:

$$\text{reactants} \underset{\text{fewer on left}}{\overset{\text{fewer on right}}{\rightleftharpoons}} \text{products}$$

How equilibria respond to temperature

According to le Châtelier's principle, an increase in the temperature of an equilibrium mixture shifts the position of equilibrium in the direction that tends to minimize the increase by absorbing energy.

Suppose the reaction is endothermic (ΔH positive), then an increase in the temperature shifts the equilibrium towards products, because that results in the absorption of energy and tends to reduce the temperature rise. On the other hand, if the reaction is exothermic, an increase in temperature favours the reactants, because their formation involves an absorption of energy and therefore tends to reduce the rise of temperature.

We can summarize this behaviour as follows:

Endothermic reactions: a rise in temperature favours products (K increases).
Exothermic reactions: a rise in temperature favours reactants (K decreases).

The $N_2O_4 \rightleftharpoons 2NO_2$ reaction is endothermic in the forward reaction (like all dissociations, it requires energy). Therefore we predict that on increasing the temperature the dissociation equilibrium will shift to the right in favour of nitrogen dioxide. In practice this means that the colour of the sample becomes more intense as the temperature is raised, because there is then a higher proportion of the brown NO_2 molecules. This is observed.

The ammonia synthesis is exothermic ($\Delta H_m^\ominus = -92.2 \text{ kJ mol}^{-1}$ at 25 °C). Without any calculation, we can therefore predict that the equilibrium shifts in favour of the *reactants* when the temperature is raised. It follows that the yield from an ammonia synthesis plant can be maximized by working at low temperatures (and, as we saw in the preceding section, at high pressures). This information immediately presents a problem: the synthesis reaction is very slow at low temperatures, and so although the equilibrium then favours the product, it is formed extremely slowly. This is typical of the problems that are encountered in real life, but there is a way round it. The solution is to use a *catalyst* (a substance that speeds up a chemical reaction without being used up in the process; we deal with them in more detail in Chapter 14). *Catalysts do not affect the position of equilibrium, but they do affect the rate at which it is attained.* In the original Haber-Bosch process the catalyst was iron but now iron together with a mixture of metal oxides is normally used. The synthesis is run at a moderate temperature (525 °C, 800 K) so that the rate is acceptable and the equilibrium position tolerably in favour of ammonia, and at 200 atm, which favours ammonia production but is not so high as to require very expensive high-pressure plant.

10.3 Equilibria between phases

Closely related to chemical equilibria are *distribution equilibria*, in which a substance dissolves to different extents in two or more phases. These can also be expressed in terms of le Châtelier's principle and, for quantitative predictions, in terms of an equilibrium constant.

Gases in liquids

Consider the distribution of a gas between a solvent and the gas phase itself. We limit attention to cases where the gas does not react with the solvent, and so while we shall include oxygen dissolving in water or in animal fat, we exclude its dissolution in blood. Gas solubilities usually decrease with increasing temperature; at higher temperatures the solvent in effect shakes the gas out of solution. Helium has a very low solubility; that is one of the reasons why it is used instead of nitrogen in deep sea diving, for less dissolves in the blood stream and the likelihood of the 'bends' is minimized. Oxygen is more soluble in water than is nitrogen, because it becomes attached to the water molecules through hydrogen bonds.

The mass of gas that dissolves depends on its partial pressure. Le Châtelier's principle predicts that applied pressure should push the gas into solution. The behaviour observed for sparingly soluble gases is summarized by a law formulated by Henry in the earliest days of physical chemistry; *Henry's law* (1803) states that *the mass of gas dissolved by unit mass of solvent is proportional to its partial pressure:*

Henry's law: $M = K_H p$. (10.3.1)

Table 10.2 Henry's law constants at 25 °C, $K_H/g\,kg^{-1}\,Pa^{-1}$

	Water	Benzene
N_2	1.80×10^{-7}	1.51×10^{-6}
O_2	4.03×10^{-7}	—
CO_2	1.46×10^{-5}	4.93×10^{-5}

Principles of physical chemistry

K_H, a kind of equilibrium constant, depends on both the gas and the solvent. Once values of K_H have been measured, Table 10.2, they are useful for discussing the dissolution of air and other gases in ponds, and in fluids such as blood plasma (as distinct from blood itself).

Partition equilibria and chromatography

When a solid is shaken with two immiscible liquids it dissolves in them to different extents. At equilibrium, which is called a *partition equilibrium*, the concentration of the solute may be $[B]_1$ in one liquid and $[B]_2$ in the other. The ratio of concentrations at equilibrium is called the *distribution coefficient* for the system:

$$K_{dist} = \{[B]_1/[B]_2\}_{eq}. \tag{10.3.2}$$

Some distribution coefficients are listed in Table 10.3. The data show that polar solutes favour polar solvents. The distribution coefficient reflects the energetics of the dissolution – energetically favourable interactions weight the partition equilibrium in favour of that solvent.

Table 10.3 Distribution coefficients at 25 °C between water (1) and another solvent (2)

Solute B	Solvent 2	$K_{dist} = \{[B]_1/[B]_2\}_{eq}$
Cl_2	CCl_4	0.10
I_2	CCl_4	0.012
$UO_2(NO_3)_2$	$(C_2H_5)_2O$	1.2
CH_3COOH	C_6H_6	16.0
$CH_2ClCOOH$	C_6H_6	28.0

Example: Iodine was extracted from an aqueous solution by shaking with one third its volume of tetrachloromethane. What proportion of the iodine will be present in the organic solvent?

Method: The distribution coefficient refers to the ratio of concentrations. Let the total amount of iodine be n, then the amounts at equilibrium will be $(1-\alpha)n$ in water (1) and αn in the organic solvent (2). The volumes are V and $\frac{1}{3}V$, respectively. The concentrations at equilibrium are therefore $(1-\alpha)n/V$ and $3\alpha n/V$. Insert these values into eqn (10.3.2) and solve for α. Refer to Table 10.3 for $K_{dist} = 0.012$.

Answer: The equilibrium is described by

$$K_{dist} = \left\{\frac{[I_2]_1}{[I_2]_2}\right\}_{eq} = \frac{(1-\alpha)n/V}{3\alpha n/V} = \frac{1-\alpha}{3\alpha}.$$

This rearranges to

$$\alpha = \frac{1}{1 + 3K_{dist}} = \frac{1}{1 + 0.036} = 0.97.$$

Therefore, 97% of the iodine is in the tetrachloromethane.

Comment: Oxidizing agents are often detected by adding acidified potassium iodide and looking for the formation of iodine by extracting it with tetrachloromethane, as in this *Example*. Notice the efficiency of the solvent extraction procedure; this is on account of the favourable van der Waals interactions between the solvent and the iodine.

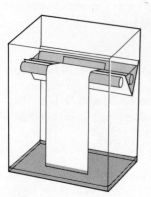

10.5 Paper chromatography

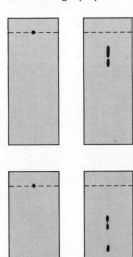

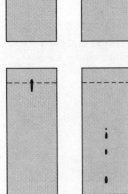

10.6 Six stages in the separation of a mixture by chromatography

The important technique of *chromatography* depends on the partition equilibrium between two phases, one stationary, the other moving. The different distribution coefficients of species are then exploited in order to separate and identify the components of a mixture.

In *paper chromatography* the stationary phase is water held on paper (that is, soaked paper), Fig. 10.5, and the moving phase is some solvent; the latter runs down the paper and the solutes are partitioned between it and the water, Fig. 10.6. The solutes separate out according to their distribution coefficients, those that favour the moving phase being carried down the paper more quickly. The paper acts only as a support for the stationary phase, and is often replaced by some other medium, such as a column filled with aluminium oxide or a similar material. *Thin-layer chromatography* is an adaptation in which the stationary phase is trapped in a thin layer of the support medium (frequently aluminium oxide) spread on a glass or celluloid film.

In *gas–liquid chromatography* (GLC), the stationary phase is a liquid trapped on some finely divided inert support held in a long thin tube, Fig. 10.7. The moving phase is a gas. The sample to be analysed is injected into the gas stream just before it enters the stationary phase, and is washed through by the stream. The components of the mixture are partitioned to different extents between the two phases, and so emerge at different times. In sophisticated instruments the detector on the outlet tube is a mass spectrometer, and the sample can be analysed directly and almost automatically. The technique is exceedingly sensitive, and is used to detect small quantities of alcohol in blood and urine, or small quantities of explosives, and to separate and identify mixtures of similar species. Figure 10.8, for example, shows the *gas-chromatograms* of lemon oil and lime oil; the small differences of composition account for the subtle shift of odour between the two oils. Whisky has been analysed by gas chromatography, Fig. 10.9; but mixing the appropriate chemicals has not yet, for some regrettable reason, reproduced the taste.

Some synthetic materials, the *ion-exchange resins*, retain different ions for different times. When they are used in an *ion-exchange chromatography* column they can separate similar species. Simply by washing the

Principles of physical chemistry

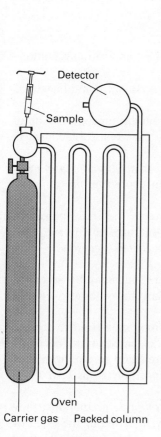

Detector

Sample

Carrier gas   Packed column

Oven

10.7 Gas–liquid chromatography

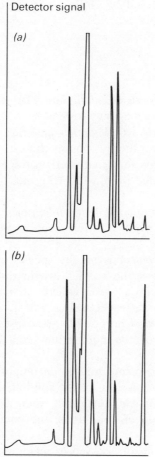

Detector signal

(a)

(b)

← Elution time ——

10.8 The gas-chromatograms of (a) lemon oil, (b) lime oil

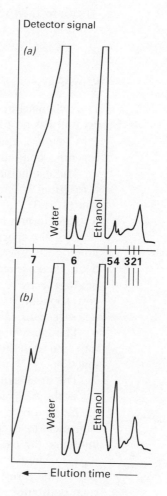

Detector signal

(a)

Water

Ethanol

7   6   54 321

(b)

Water

Ethanol

← Elution time ——

10.9 The gas-chromatograms of (a) scotch whisky, (b) bourbon whiskey. (Key: 1, ethanal; 2, methanal; 3, ethyl methanoate; 4, ethyl ethanoate; 5, methanol; 6, propan-1-ol; 7,3-methylbutan-1-ol)

sample through the column with water, the different species emerge at different times and may be collected separately. The time for a species to be washed through a column is called its *elution time*; the washing process itself is *elution*. This technique is especially useful for the separation of the rare-earth (lanthanide) elements and of the transuranium elements. An everyday application is in water softeners, where the ion-exchange resin (often an artificial zeolite, a complicated alumino-silicate with an open structure) captures calcium ions and replace them with sodium ions.

They are then recharged by elution with concentrated sodium chloride solution.

Summary

1   A chemical reaction is at *equilibrium* when the composition is stationary and has no further tendency to react.

2   Chemical equilibria are *dynamic equilibria*; the forward and backward reactions continue, but there is no net change.

3   The *equilibrium constant* characterizes the equilibrium composition of a reaction. It is defined in eqn (10.1.2) and, for gas-phase reactions, in eqn (10.1.3).

4   Equilibrium constants are *independent of the concentrations and pressures* of the species involved in the reaction, but *depend on the temperature.*

5   In *homogeneous reactions* all the species are in the same phase; in *heterogeneous reactions* the species are in more than one phase. The corresponding equilibria are called *homogeneous equilibria* and *heterogeneous equilibria,* respectively.

6   The *concentrations of pure solids* are constant, and may be absorbed into the definition of the equilibrium constant for a heterogeneous reaction.

7   The equilibrium constant for the decomposition of a solid to form a gas is the *dissociation pressure* of the solid at the temperature specified.

8   *Le Châtelier's principle* states that when a reaction at equilibrium is subjected to a change of conditions, the composition adjusts so as to minimize the change.

9   Le Châtelier's principle is used to make *qualitative* predictions; they can be interpreted and expressed *quantitatively* in terms of the equilibrium constant.

10   If the *concentration* of a species involved in an equilibrium is artificially increased, the reaction adjusts the equilibrium compositions so as to minimize the increase.

11   If a reaction involving gases is subjected to an *increase in pressure* the reaction adjusts the composition so as to reduce the number of gas-phase molecules (and thereby to minimize the increase in pressure).

12   Equilibria that do not involve gases are insensitive to moderate changes of pressure.

13   An *increase of temperature* shifts the position of equilibrium in favour of products in the case of endothermic reactions, and in favour of reactants for exothermic reactions.

14   The yield in the ammonia synthesis is increased by *increasing the pressure* and *decreasing the temperature.* Decreasing the temperature also has the result of slowing the rate at which equilibrium is reached. Therefore a catalyst is added.

15   The *presence of catalysts* affects the rates of reactions, not their positions of equilibrium.

16   *Henry's law* states that the mass of gas dissolved is proportional to

Principles of physical chemistry

its partial pressure, eqn (10.3.1).

17 The *distribution coefficient* expresses the *partition equilibrium* between two phases, eqn (10.3.2).

18 *Chromatography* (paper, thin-layer, gas–liquid, and ion-exchange) is a technique for separating the components of a mixture making use of their different partition equilibria between moving and stationary phases.

Problems

1 Explain the nature of chemical equilibrium and describe the evidence showing that it is dynamic.

2 Write expressions for the equilibrium constants for the following reactions: (a) $CdCO_3(s) \rightleftharpoons CdO(s) + CO_2(g)$, (b) $2SO_2(g) + O_2(g) \rightleftharpoons 2SO_3(g)$, (c) $2SO_3(g) \rightleftharpoons 2SO_2(g) + O_2(g)$, (d) $2H_2(g) + O_2(g) \rightleftharpoons 2H_2O(l)$, (e) $C_2H_5COOH(l) + CH_3OH(l) \rightleftharpoons C_2H_5COOCH_3(l) + H_2O(l)$.

3 Explain why the concentration of a pure solid is a constant.

4 Explain why the equilibrium constant for the decomposition of a pure solid to a gas is equal to the decomposition pressure of the solid.

5 The equilibrium constant for the decomposition of cadmium carbonate, reaction (a) in Question 2, has the value 0.5 kPa at 550 K and 25 kPa at 600 K. Describe what happens when a sample of the solid is heated to 600 K in a vessel fitted with a piston exerting a pressure of 10 kPa, but which cannot be pushed out beyond some point, and then allowed to cool back to room temperature.

6 State le Châtelier's principle. Why is it limited to states of dynamic equilibrium and does not apply to static equilibria (*e.g.*, a pencil balanced on its point)?

7 State what direction is favoured when the reactions in Question 2 are subjected to an increase of pressure.

8 Is the water synthesis equilibrium shifted in favour of products by an increase of temperature?

9 An inventor files a patent for a substance he calls *Anticat* which, he claims, eliminates the tendency of hydrogen and oxygen to combine to form water. Comment on the plausibility of the claim. In what sense might he be correct, and how should he reword his claim?

10 A steel vessel containing ammonia, hydrogen, and nitrogen is at equilibrium at 1000 K. Analysis of the contents shows that the concentrations are $[NH_3] = 0.142 \text{ mol dm}^{-3}$, $[H_2] = 1.84 \text{ mol dm}^{-3}$, and $[N_2] = 1.36 \text{ mol dm}^{-3}$. Calculate the value of K_c for the reaction $N_2(g) + 3H_2(g) \rightleftharpoons 2NH_3(g)$ at this temperature.

11 Calculate the value of K_p for the reaction in the last Question on the assumption that the gases behave perfectly.

12 The reaction $2H_2S(g) \rightleftharpoons 2H_2(g) + S_2(g)$ has $K_c = 1.06 \times 10^{-6} \text{ mol dm}^{-3}$ at 750 °C. What amount of substance of $S_2(g)$ (expressed as moles of S_2) will be present at equilibrium in a 5.00 dm^3 vessel in which there are known to be 2.21 mol $H_2S(g)$ and 1.17 mol $H_2(g)$?

13 At 2200 K and a total pressure of 1 atm, steam is 1.18% dissociated into hydrogen and oxygen. Calculate K_p for the reaction $2H_2O(g) \rightleftharpoons 2H_2(g) + O_2(g)$.

14 When molecular iodine is dissolved in an aqueous solution of potassium iodide, the following equilibrium is established:

$$I_3^- \rightleftharpoons I_2 + I^-$$

[The equilibrium constant is 0.0015 mol dm^{-3} at 298 K.]

0.02 mole of iodine is dissolved in 500 cm^3 of a solution containing

0.2 mol dm^{-3} of potassium iodide at 298 K. Calculate the concentration of each ion present at equilibrium.

The above equilibrium may be investigated by partition between two solvents, a method which can sometimes be used to determine the molecularity of a compound in solution. Benzene carboxylic acid (benzoic acid) is more soluble in benzene than in water. In benzene it exists as double molecules. Explain how you could confirm this fact experimentally using a partition method, giving important experimental details and the relevant theoretical background.

(Welsh S)

15   Write a *short account* of the factors affecting the position of equilibrium of a balanced reaction, the rate at which equilibrium is attained and the value of the equilibrium constant.

(a) Using partial pressures, show that for gaseous reactions of the type

$$XY(g) \rightleftharpoons X(g) + Y(g)$$

at a given temperature, the pressure at which XY is exactly *one-third* dissociated is numerically equal to *eight* times the equilibrium constant at that temperature.

(b) When one mole of ethanoic acid (acetic acid) is maintained at 25 °C with 1 mole of ethanol, one-third of the ethanoic acid remains when equilibrium is attained. How much would have remained if one-half of a mole of *ethanol* had been used instead of one mole at the same temperature?

(SUJB)

16   This question concerns the reaction of nitrogen and hydrogen to form ammonia which has an equilibrium constant (K_p) value of $5 \times 10^5 \text{ atm}^{-2}$ at 25 °C and one atmosphere pressure. Under these conditions the bond energies [enthalpies] (in kJ mol^{-1}) are $N \equiv N$, 945; $H—H$, 436; $N—H$, 391.

In the Haber-Bosch industrial process, gaseous nitrogen and hydrogen in a mole ratio of $1:3$ are reacted, typically at a temperature of 500 °C, at a pressure of 200 atm and in the presence of finely divided iron when the equilibrium mole percentage of ammonia in the product mixture is 18.

(a) Give the equation for the reaction.

(b) Calculate a value for the standard enthalpy change for the reaction $(\Delta H^{\ominus})$ at 25 °C.

(c) Give an expression for K_p in terms of the partial pressures of the gases.

(d)   (i) Comment on the effect of increasing the pressure on the equilibrium constant.

(ii) Explain why high pressures are used in the industrial process.

(iii) Explain whether the high pressures in the industrial process could be achieved by adding an excess of an inert gas (*e.g.*, argon).

(e) Outline the reasons for using a temperature of 500 °C for the industrial process.

(f) Explain the use of 'finely divided iron'.

(g)   Calculate the mole percentage conversion of nitrogen to ammonia for the industrial process.

(Welsh)

17   (a) State one essential characteristic of a chemical equilibrium.

(b) State two ways in which the time taken to establish such an equilibrium may be altered.

(c) The following data indicate the effect of temperature and pressure on the equilibrium concentration of the product, X, of the forward reaction in a gaseous equilibrium.

	Percentage of X present in the equilibrium mixture at		
Temperature/°C	1 atm	100 atm	200 atm
550	0.077	6.70	11.9
650	0.032	3.02	5.71
750	0.016	1.54	2.99
850	0.009	0.87	1.68

(i) Use the above data to deduce whether the production of X is accompanied by an increase or a decrease in volume and explain your answer.

(ii) Use the above data to determine whether the production of X is an exothermic or an endothermic process and explain your answer.

(iii) State qualitatively the theoretical optimum conditions of temperature and pressure suggested by the above data for the commercial production of X.

(JMB)

18 The distribution (partition) coefficient of a compound X between trichloromethane and water is 12, X being more soluble in trichloromethane than in water. What mass of X will be extracted from $200 \, cm^3$ of an aqueous solution containing 8 g of X by shaking it with $50 \, cm^3$ of trichloromethane?

(Oxford)

19 Discuss briefly the principles of chromatography.

Give brief details of the experimental methods used for column chromatography, thin-layer chromatography and paper chromatography giving a specific example of the use of each specialized technique.

Suggest analytical techniques (not necessarily chromatographic) which would be most appropriate for quantitative analysis of the major components of two of the following samples:

(a) a mixture of nitrogen(II) oxide and ethane,
(b) a sample of North Sea oil,
(c) a mineral sample *brought back* to Earth from the Moon,
(d) a mineral sample *on* the planet Mars.

(Oxford and Cambridge S)

20 If lead(II) chloride is precipitated in the presence of thorium nitrate, using an aqueous solution of lead(II) nitrate and dilute hydrochloric acid, the lead(II) chloride contains radioactive ^{212}Pb atoms. (These radioactive lead atoms are a daughter product of the thorium.)

Show how you could experimentally use this information to establish that lead(II) chloride and its saturated solution are in dynamic equilibrium.

(SUJB; continued from qu. 1.15)

21 When a substance is added to a two-phase liquid system it is generally distributed with different equilibrium concentrations in the two phases. Why is this so?

A certain amount of iodine was shaken with CS_2 and an aqueous solution containing $0.3 \, \text{mol dm}^{-3}$ of potassium iodide. By titrating with thiosulphate solution, it was found that the CS_2 phase contained $32.3 \, \text{g dm}^{-3}$ and the aqueous phase $1.14 \, \text{g dm}^{-3}$ of iodine. The distribution coefficient for iodine distributed between CS_2 and water is 585. Calculate the equilibrium constant for the reaction $I_2 + I^- \rightleftharpoons I_3^-$ at the prevailing temperature.

(Cambridge Entrance)

22   Discuss the importance of the concept of equilibrium constant in chemistry, including examples from as wide a range of applications as possible.

(Cambridge Entrance)

The natural direction of change: the Second law

We shall see that it is possible to account for all natural processes in terms of a single idea, and that in order to express it precisely we are led to the concept of entropy. We shall also see that for chemical applications it is sensible to introduce a related quantity, the Gibbs function. The latter is extremely important because it lets us predict the values of equilibrium constants. Moreover, it also lets us judge the energy resources of chemical reactions taking place in biological systems and in sources of electrical energy, such as fuel cells.

Introduction

So far we have treated the equilibrium constant only as something measured experimentally. In this chapter, the world of the *Second law of thermodynamics*, we look into what determines its value. The Second law deals with the direction of natural change, questions such as why a reaction runs in one direction and not another, and the position of equilibrium.

We shall look very briefly at two aspects of the Second law. First we shall see what governs the natural direction of change. The Second law has such a simple interpretation that, even without going into the thermodynamic calculations themselves, we are led to the heart of understanding why processes take place in one direction and not another. The second aspect is of great practical utility. The Second law lets us construct tables of data that can be used to predict the values of equilibrium constants, to decide how they depend on temperature, and to assess the quantity of electrical energy available from reactions.

11.1 Why things change

We all know that some things happen naturally while others do not. A hot block of metal grows cool, Fig. 11.1; a cool block does not spontaneously grow hot. Iron rusts, Fig. 11.2; rust does not spontaneously decay into iron and release oxygen. What is the reason?

The dispersal of energy

There is a feature common to all types of natural change: *they are all*

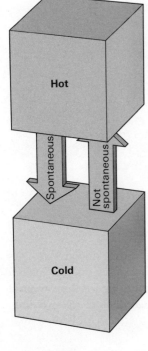

11.1 Cooling is a spontaneous process

accompanied by an increase in the dispersal of energy. At this stage, take dispersal to have its everyday meaning. Energy that is not dispersed can be thought of as energy confined to a small, well-defined location. Energy that has dispersed can be thought of as being spread more widely, such as into the motion of the particles in the surroundings of some object.

As a first example of how dispersal governs change, take the case of a hot block of metal in contact with cooler surroundings (your hand or the atmosphere). The block is a collection of electrons and vigorously vibrating ions, Fig. 11.3(*a*). At this stage the energy is concentrated within the volume occupied by the block. The ions in the block jostle their neighbours, and those at the edge jostle the particles in the surroundings. The latter pick up energy as a result of the jostling, and in turn pass it on to their neighbours, Fig. 11.3(*b*). There are very many particles in the surroundings, and so the energy leaks away. The natural direction of change is in the direction of the dispersal of energy. In this case it corresponds to the cooling of the block to the temperature of its surroundings. The fact that we never see a block in a cool room becoming hot spontaneously is because it is extremely improbable that by random jostling a lot of energy in the surroundings will return to the block all at the same time.

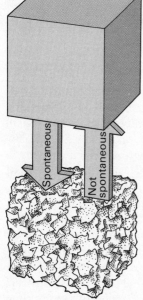

11.2 Rusting is a spontaneous process

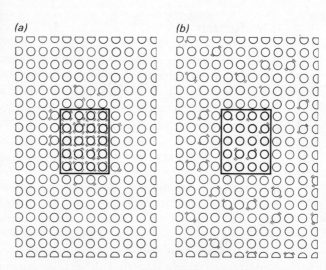

11.3 Energy initially localized (*a*) tends to disperse (*b*)

Every type of natural change can be explained along similar lines. As chemists we are interested in the natural direction of reactions. Since a reaction changes the nature of the substances, accounting for their natural direction is trickier than in the case of a physical change like cooling. Nevertheless it can be done, as we now see.

Chemical reactions as dispersers of energy
Consider the changes that occur when iron rusts. A sufficiently close

model reaction is:

$$4Fe(s) + 3O_2(g) \rightarrow 2Fe_2O_3(s); \quad \Delta H_m^\ominus = -1648.4 \text{ kJ mol}^{-1} \quad \text{at } 25\,^\circ C.$$

The oxidation is exothermic, and so energy is released when it takes place. This released energy disperses into the surroundings, Fig. 11.4. The natural direction of change in the reaction is from the reactants, with the energy locked up in the bonds holding the metal together and in the molecules of oxygen, to the product, where some energy remains in the highly localized solid (the heap of rust) but 1648 kJ for every 2 mol Fe_2O_3 has jostled away and is spread over the surroundings.

Once we stop to think about the reaction we see that there is more to it than this discussion suggests. For instance, the oxygen is initially present as a gas and so its energy is spread over a large volume (*e.g.*, spread over about 72 dm³ if we are thinking of 3 mol O_2 at 1 atm pressure and 25 °C). When the reaction has taken place the products are entirely solid, and 2 mol Fe_2O_3 occupies only about 0.06 dm³: all the energy possessed by the compound is now confined to that volume, Fig. 11.5. We see that there is some kind of competition between the *localization* of energy that occurs when the widely dispersed oxygen gas reacts to produce a highly localized solid and the *dispersal* of the energy that is released into the surroundings as a result of the reaction.

Assessing the competition between the localization of energy by the elimination of a gas and the dispersal of the liberated reaction energy looks as though it could be quite involved. In fact it is simple once we have introduced another and very famous thermodynamic quantity, the entropy.

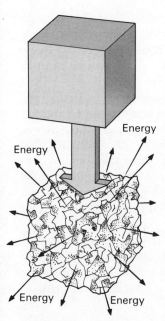

11.4 The (exothermic) rusting of iron disperses energy into the surroundings

11.2 Entropy and the Second law

The formal definition of entropy is set out in the Appendix, p. 178. For our purposes its interpretation is what is important. *The entropy is a measure of the chaotic dispersal of energy.* As the dispersal of energy increases, so too does the entropy. Whenever a system gets more chaotic, like a pack of cards being shuffled, then the entropy increases. We now develop this idea.

The Second law of thermodynamics

We have already seen that the direction of natural change is towards increasing dispersal. In terms of the entropy, natural change corresponds to the direction of increasing entropy. This statement is nothing other than the Second law of thermodynamics (but that law was built up on the basis of much more rigorous arguments than we have expressed here):

> *Second law of thermodynamics: The entropy of the universe increases in the course of every natural change.*

Bear in mind in the course of reading this chapter the present concern about the world's energy resources. It should be clear that the First law eliminates the conventional worry; if energy cannot be destroyed, there is

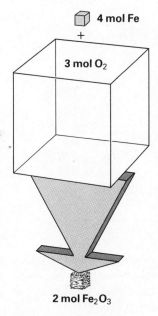

4 mol Fe

+

3 mol O_2

2 mol Fe_2O_3

11.5 The reactants become more localized

no need to worry about using it up. It will become clear that it is the *quality* of the energy (in the sense of its dispersal) which should be the true object of concern. Mankind is conserving energy but producing too much entropy and destroying energy's availability. We have an entropy crisis, not an energy crisis.

Before we go on there are two words of warning. The Second law refers in a grand way to the 'universe'. In practice that means the contents of the reaction vessel and its immediate surroundings, such as the atmosphere or a water-bath. The first warning therefore is that the calculation of the direction of natural change must take into account the entropies both of the system itself *and* of its surroundings. The entropy of the system itself may drop, but the corresponding process will still be a natural, spontaneous change if that drop is compensated by a larger increase of entropy in the surroundings.

The other warning concerns the meaning of 'spontaneous'. Spontaneous does not mean fast. Spontaneous means 'able to occur without needing work to bring it about'. Always remember that thermodynamics refers to the natural *direction* of change, and not to the *rate* of change.

Entropies of substances

Entropy is always denoted by the letter S, and molar entropy, the entropy of unit amount of substance, by S_m. *Standard molar entropies*, the entropies of unit amounts of substance in their standard states at the temperature specified, are denoted $S_m^{\ominus}$. The precise thermodynamic definition of the entropy (Appendix, p. 178) leads to a way of measuring entropies by a straightforward laboratory experiment using a calorimeter, and so tables of values, such as Table 11.1, can be compiled.

Table 11.1 Standard molar entropies at 25 °C, $S_m^{\ominus}/J\,K^{-1}\,mol^{-1}$

Solids		Liquids		Gases	
C(graphite)	5.7	Hg	76.0	H_2	130.6
C(diamond)	2.4	H_2O	69.9	N_2	192.1
Fe	27.3			O_2	205.0
Cu	33.1			CO_2	213.6
AgCl	96.2	C_2H_5OH	160.7	NO_2	239.9
Fe_2O_3	87.4	C_6H_6	173.3	N_2O_4	304.0
$CuSO_4 \cdot 5H_2O$	300.4	CH_3COOH	159.8	NH_3	192.3
Sucrose	360.2			CH_4	186.2

The first point to notice about the values in Table 11.1 are the units. Entropies are normally expressed as so many joules per kelvin, $J\,K^{-1}$, and so molar entropies are normally expressed in $J\,K^{-1}\,mol^{-1}$. (The molar entropy has the same dimensions as the molar gas constant R.)

The second point to notice is the relative magnitudes of the entries in the table. The molar entropies of gases at 1 atm pressure and 25 °C are all roughly the same, and are much larger than the molar entropies of most solids. This is because in a gas the energy is due to the motion of the particles, and the particles themselves (with their energies) are dispersed over a large volume. On the other hand, the energy of a solid is confined to a small region, the volume it occupies, and the particles can only vibrate. Note, however, that the entropies of solids composed of complex molecules, such as sucrose, can be very high. This is because the molecules have many atoms and the energy can be shared among them.

We can also see from Table 11.1 that the molar entropies of liquids are intermediate between those of gases and solids. This is in line with their intermediate structures. Notice how the molar entropy of water is quite low in comparison with other liquids at the same temperature. This is yet another way in which hydrogen bonds reveal themselves. In this case they are holding the molecules together into a fairly uniform structure, and water is more 'solid-like' in its structure than are other liquids.

Changes of entropy

The change of entropy when reactants change completely into products can be calculated very simply by taking the appropriate combinations of the entropies listed in Table 11.1. Note, however, that the values listed there refer to the standard states of all the components, and so for a reaction such as the rusting of iron the entropy change obtained from the data is the value when oxygen gas at 1 atm pressure combines with solid iron and reacts completely to form rust. Likewise, if we use the values given there in the case of the formation of ammonia, the value of the entropy change corresponds to the conversion of pure hydrogen at 1 atm and pure nitrogen at 1 atm to pure ammonia at 1 atm.

As an example of the use of the data, consider the oxidation of iron at 25 °C. The molar entropy change accompanying the reaction $4Fe(s) + 3O_2(g) \rightarrow 2Fe_2O_3(s)$, with all species in their standard states, is the *standard molar entropy of reaction* $\Delta S_m^\ominus$:

$$\Delta S_m^\ominus = 2(87.4 \text{ J K}^{-1} \text{ mol}^{-1}) - 4(27.3 \text{ J K}^{-1} \text{ mol}^{-1})$$
$$- 3(205.0 \text{ J K}^{-1} \text{ mol}^{-1})$$
$$= -549.4 \text{ J K}^{-1} \text{ mol}^{-1}.$$

As anticipated, there is a large *decrease* in entropy in this reaction because the highly dispersed oxygen gas reacts to form a compact solid. The interesting point about this result is that we have been able to attach a *numerical* value to the reduction in dispersal brought about by the reaction.

Why then is rusting spontaneous? The calculation seems to show that it is accompanied by a *decrease* in entropy. The missing feature is that we have not yet taken into account the change of entropy in the surroundings. This emphasizes how important it is to consider both contributions to the change of entropy. We return to this point shortly.

The same technique can be used to find the change of entropy during the complete dissociation of N_2O_4 at 25 °C and 1 atm pressure. We can expect the entropy to *increase* because there are 2 mol NO_2 for every

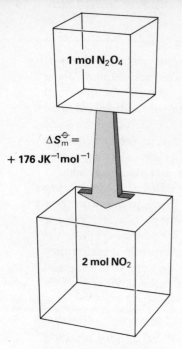

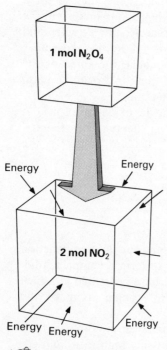

11.6 The entropy change of the system

$\Delta S_m^{\ominus} =$
$+ 176\ \mathrm{JK^{-1}mol^{-1}}$

$\Delta S_m^{\ominus}$ (surroundings) $=$
$-192\ \mathrm{JK^{-1}\ mol^{-1}}$

11.7 The entropy change of the surroundings during an endothermic reaction

1 mol N_2O_4, and so at constant pressure complete reaction doubles the volume over which the energy of the species is dispersed, Fig. 11.6. Application of the data in Table 11.1 gives a value of $+175.8\ \mathrm{J\ K^{-1}\ mol^{-1}}$ for the standard molar entropy of the reaction $N_2O_4(g) \rightarrow 2NO_2(g)$, in agreement with this prediction.

Changes of entropy in the surroundings

Before we can make use of the information just derived we have to be able to assess the change of entropy in the surroundings of the reaction vessel. This turns out to be very easy because thermodynamic arguments show that it can be expressed in terms of the enthalpy of reaction. If the enthalpy of reaction is ΔH, and the temperature is T, then the change of entropy of the surroundings is given by

$$\Delta S(\text{surroundings}) = -\Delta H/T. \tag{11.2.1}$$

(The T in the denominator comes from the thermodynamic definition of entropy, Appendix, p. 178.) This formula agrees with common sense about the way that reactions affect the dispersal of energy. Suppose the reaction is exothermic. Intuitively we then expect the energy it releases into the surroundings to increase their entropy. This is also the result predicted by eqn (11.2.1) because in the case of an exothermic reaction ΔH is negative, and so $\Delta S(\text{surroundings})$ is positive, corresponding to an increase from its initial value.

The changes of entropy in the surroundings accompanying a chemical reaction can be obtained very easily from the information about reaction enthalpies discussed in Chapter 9 and in particular the data in Table 9.3. For instance, the standard molar reaction enthalpy of the iron oxidation reaction is $-1648.4\ \mathrm{kJ\ mol^{-1}}$ and so the standard molar entropy change of the surroundings at 25 °C (298.15 K) is

$$\Delta S_m^{\ominus}(\text{surroundings}) = \frac{-(-1648.4 \times 10^3\ \mathrm{J\ mol^{-1}})}{(298.15\ \mathrm{K})} = +5529\ \mathrm{J\ K^{-1}\ mol^{-1}}.$$

This is a massive *increase* in entropy, and arises because the reaction releases energy into the surroundings.

The dissociation of N_2O_4 is endothermic and $\Delta H_m^{\ominus} = +57.2\ \mathrm{kJ\ mol^{-1}}$. The entropy change in the surroundings, Fig. 11.7, is therefore

$$\Delta S_m^{\ominus}(\text{surroundings}) = \frac{-(+57.2\ \mathrm{kJ\ mol^{-1}})}{(298.15\ \mathrm{K})} = -192\ \mathrm{J\ K^{-1}\ mol^{-1}}.$$

All endothermic reactions decrease the entropy of the surroundings.

Total entropy changes

The total change of entropy is the sum of the entropy changes that take place in the reaction vessel and in the surroundings. Therefore we can write

$$\Delta S(\text{total}) = \Delta S(\text{surroundings}) + \Delta S(\text{reaction vessel})$$
$$= -\Delta H/T + \Delta S(\text{reaction vessel}). \tag{11.2.2}$$

Principles of physical chemistry

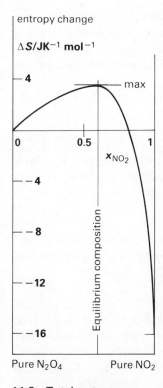

entropy change

$\Delta S/\mathrm{JK^{-1}\,mol^{-1}}$

Equilibrium composition

Pure N_2O_4 Pure NO_2

11.8 Total entropy changes during the course of reaction at 1 atm and 25 °C. The spontaneous direction is from either pure N_2O_4 or pure NO_2 to the composition corresponding to maximum total entropy. (The equilibrium composition corresponds to that depicted in Fig. 10.2, $p = 1$ atm)

From now on we shall drop the label 'reaction vessel' and remember that whenever unlabelled ΔS appears it refers to the system itself. In order to find ΔS(total) we simply have to add together the values found in the two preceding sections.

In the case of the rusting of iron, the total entropy change for the reaction when the reactants in their standard states change completely to product (rust) in its standard state is readily found to be

$$\Delta S_m^{\ominus}(\text{total}) = (+5529\ \mathrm{J\,K^{-1}\,mol^{-1}}) + (-549\ \mathrm{J\,K^{-1}\,mol^{-1}})$$
$$= +4980\ \mathrm{JK^{-1}\,mol^{-1}}.$$

That is, for every 4 mol Fe converted to 2 mol Fe_2O_3 there is an overall *increase* in the entropy of the universe of almost $5000\ \mathrm{J\,K^{-1}\,mol^{-1}}$. This is a large positive quantity, and so the reaction is spontaneous. That is the thermodynamic reason why steel corrodes. It may be possible to hinder the rate at which the corrosion takes place, and that is one of the reasons for electroplating or painting bridges, ships, and cars. Nevertheless, we can see from this result that iron (and steel) structures are fundamentally unstable and have a built-in tendency to decay.

Similarly, the total entropy change accompanying the complete dissociation of N_2O_4 at 1 atm and 25 °C is

$$\Delta S_m^{\ominus}(\text{total}) = (-191.8\ \mathrm{J\,K^{-1}\,mol^{-1}}) + (175.8\ \mathrm{J\,K^{-1}\,mol^{-1}})$$
$$= -16.0\ \mathrm{J\,K^{-1}\,mol^{-1}}.$$

This is negative and small, and so we can conclude that the reaction does not go to completion. Nevertheless, because this entropy change is so small, there may be an intermediate stage of the reaction where the entropy goes through a maximum, and so changes towards that composition, either from pure reactants or from pure products, will be spontaneous. The dependence of the *total* entropy on the composition of the reaction mixture is shown in Fig. 11.8. There is a composition corresponding to maximum entropy. Furthermore, it corresponds exactly to the equilibrium composition predicted on the basis of the equilibrium constant in Section 10.1. The same kind of calculation done for ammonia and for water would show a maximum overall entropy when the mixture had reached the equilibrium composition met in Section 10.1.

11.3 The Gibbs function

We have emphasized that whether or not a reaction has a natural tendency to occur under the specified conditions depends on the value of ΔS(total). That quantity is composed of two parts. One is the entropy change in the reaction vessel itself. The other indicates the extent of dispersal in the surroundings, and can always be calculated very simply in terms of the reaction enthalpy. The total entropy change can therefore be expressed entirely in terms of quantities related to the reaction:

$$\Delta S(\text{total}) = -\Delta H/T + \Delta S \qquad (11.3.1)$$

where ΔS is the entropy change of the reaction system (this is eqn (11.2.2)). In order to use this expression we have to refer to two tables of data, one for the entropies of species, Table 11.1, and the other for the enthalpies of formation, such as Table 9.3. It is sensible to combine the two types of data into a single table.

As a first step we reorganize eqn (11.3.1) by multiplying through by $-T$:

$$-T\Delta S(\text{total}) = \Delta H - T\Delta S.$$

The new quantity $-T\Delta S(\text{total})$ is now given a new symbol and a new name. It is called the *Gibbs function of reaction* and denoted ΔG. The last equation therefore becomes

$$\Delta G = \Delta H - T\Delta S. \tag{11.3.2}$$

This is one of the most important equations in thermodynamics. G, the *Gibbs function*, is also called the *Gibbs free energy*. The reason for this alternative name will shortly become clear.

Using the Gibbs function

The main point about the Gibbs function is that it tells us at a glance whether or not a reaction has a tendency to occur. We already know that a reaction has a tendency to take place if $\Delta S(\text{total})$ is positive. Therefore, in terms of the Gibbs function, which is proportional to $-\Delta S(\text{total})$, *a reaction has a natural tendency to occur if ΔG is negative.* For very many reactions ΔH is quite a lot bigger than $T\Delta S$, and so in these cases we can write $\Delta G \approx \Delta H$ (as a very crude approximation). Then an exothermic reaction, one corresponding to a negative ΔH, has a negative ΔG, and is therefore spontaneous. That is why we are so familiar with spontaneous exothermic reactions. An endothermic reaction (with ΔH positive) is spontaneous only if $T\Delta S$ is big enough, because only then is $\Delta H - T\Delta S$ negative. Spontaneous exothermic reactions are driven by the large quantity of entropy they generate in the surroundings; spontaneous endothermic reactions are driven by the large quantity of entropy generated in the reaction mixture itself.

Table 11.2 Standard molar Gibbs functions of formation at 25 °C, $\Delta G_{\text{f,m}}^{\ominus}/\text{kJ mol}^{-1}$

Solids		Liquids		Gases	
NaCl	−384.0	H_2O	−237.2	NH_3	−16.5
KCl	−408.3	CH_3OH	−166.4	NO_2	51.3
NH_4Cl	−203.0	C_2H_5OH	−174.1	N_2O_4	97.8
Al_2O_3	−1582.4	C_6H_6	124.3	CO_2	−394.4
Fe_2O_3	−742.2	HCl(aq)	−131.3	CH_4	−50.8

The basic information for calculating ΔG is carried in tables of *standard molar Gibbs functions of formation*, $\Delta G_{\text{f,m}}^{\ominus}$, Table 11.2. These are the values of ΔG when unit amount of substance in its standard state is formed from its elements in their standard states at the temperature specified (which is usually but not necessarily 25 °C). This is similar to the definition of standard enthalpies of formation introduced in Section 9.3. In order to find the standard molar Gibbs function of any reaction all that it is necessary to do is to take the appropriate combinations of the $\Delta G^{\ominus}_{f,m}$ just as we combine enthalpies of formation to find reaction enthalpies. This should be clear after working through the following *Example* using the information in Table 11.2.

Example: A fuel cell (a device for producing electricity directly from a chemical reaction) uses the oxidation of methanol. Calculate the standard Gibbs function of the reaction.

Method: The reaction is

$$CH_3OH(l) + \tfrac{3}{2}O_2(g) \rightarrow CO_2(g) + 2H_2O(l).$$

The Gibbs function of the reaction is calculated by taking the difference of the sums of the Gibbs functions of formation for the products and the reactants. Use the data in Table 11.2. Gibbs functions of formation of elements are zero.

Answer: The overall Gibbs function is

$$\Delta G^{\ominus}_m = \{\Delta G^{\ominus}_{f,m}(CO_2, g) + 2\Delta G^{\ominus}_{f,m}(H_2O, l)\}$$
$$- \{\Delta G^{\ominus}_{f,m}(CH_3OH, l) + \tfrac{3}{2}\Delta G^{\ominus}_{f,m}(O_2, g)\}$$
$$= \{(-394.4) + 2(-237.2)\}\,kJ\,mol^{-1} - \{(-166.4) + \tfrac{3}{2}(0)\}\,kJ\,mol^{-1}$$
$$= -702.4\,kJ\,mol^{-1}.$$

Comment: We shall go on to see that the value of ΔG gives the electrical work that may be produced by a reaction when it is coupled to some kind of electrical device, such as an electric motor. In the present case, therefore, we know that for every 32 g of methanol (1 mol CH_3OH) consumed, 702.4 kJ of electrical work can be obtained. The standard molar enthalpy of combustion of methanol is $-764\,kJ\,mol^{-1}$, and so direct conversion to electrical energy provides 92% of the energy the reaction can provide as heat.

At this stage all the Gibbs function appears to do is to reproduce the information that we worked through in a more laborious fashion in the earlier sections. Instead of arriving at the conclusion that the complete dissociation of N_2O_4 is not spontaneous, on the grounds that it is accompanied by a net decrease in the total entropy, we have now come to the same conclusion on the basis that it is accompanied by an increase in the Gibbs function! The great advantage of the Gibbs function, however, now begins to emerge. This is because thermodynamic arguments show us how to use $\Delta G^{\ominus}$ to predict the composition at which the reaction mixture has no further natural tendency to change. In other words knowing $\Delta G^{\ominus}$ we are able to predict the *equilibrium constant* for a reaction.

The connection between the standard molar Gibbs function of a reaction and its equilibrium constant comes from thermodynamics and is as follows:

$$\Delta G_m^{\ominus} = -RT \ln K. \tag{11.3.3}$$

This expression links the work of this chapter to the work of the last. On the left of the equation there is a quantity that we can obtain from Table 11.2 by taking appropriate combinations of the entries. On the right we have the equilibrium constant, the quantity at the centre of the stage in the last chapter. Therefore we now have a way of predicting the equilibrium properties of a reaction even if it has not been investigated experimentally. The way this can be used to discuss equilibrium constants is illustrated in the following *Example*. We shall also see other examples in the following two chapters.

Example: Predict the value of the equilibrium constant for the reaction $C(s) + H_2O(g) \rightleftharpoons CO(g) + H_2(g)$ on the basis that at 1000 K the standard molar Gibbs function of the reaction is -7.99 kJ mol^{-1}.

Method: Use eqn (11.3.3) in the form $\ln K = -\Delta G_m^{\ominus}/RT$. The specification of the equilibrium constant is

$$K_p = \left\{ \frac{p(CO)p(H_2)}{p(H_2O)} \right\}_{eq},$$

its dimensions are therefore those of pressure (*e.g.*, atm).

Answer: At 1000 K the value of RT is

$$RT = (8.314 \text{ J K}^{-1} \text{ mol}^{-1}) \times (1000 \text{ K}) = 8.314 \text{ kJ mol}^{-1}.$$

Therefore the equilibrium constant has the value

$$\ln K = \frac{-(-7.99 \text{ kJ mol}^{-1})}{(8.314 \text{ kJ mol}^{-1})} = 0.961, \text{ implying } K = 2.61 \text{ atm.}$$

Comment: The reaction is part of the *water gas reaction*, which is used in the industrial reduction of steam to hydrogen. Hydrogen is used widely commercially, such as in the Haber-Bosch process for the synthesis of ammonia, and hence of nitrogen fertilizers.

Work and the Gibbs function

There is one more aspect of ΔG that opens up even wider applications than the prediction of equilibrium constants; in some ways this other aspect is even more important because it is the basis of the application of physical chemistry to biological systems, fuel cells, and electrochemistry.

Thermodynamic arguments lead to the result that *the change in Gibbs function is equal to the maximum quantity of electrical work that can be obtained by harnessing the process.* That is:

$$\Delta G = w_{\text{e,max}}. \tag{11.3.4}$$

(This relation is the basis of the name 'free energy' for G.) The great importance of this result should be obvious. In the first place, electrical cells and fuel cells, Chapter 13, are producers of electricity, and as such they are sources of electrical work. Therefore, if we know the Gibbs function for the reaction going on inside them, we can state the maximum quantity of electrical work that can be obtained. Moreover, the processes going on inside our bodies, such as thinking, are basically electrical, and so when we assess the energy resources of molecules involved in metabolism we should really do so in terms of the Gibbs function.

Once again, take the oxidation of iron. The oxidation of 4 mol Fe corresponds to a change of Gibbs function of about -1485 kJ (see Table 11.2). Therefore, if we can devise a way of tapping that energy electrically, we can produce 1485 kJ of electrical work for every 223 g of iron consumed. In some sense the world is wasting a vast quantity of energy simply by letting steel objects corrode; it is like burning the iron and discarding the energy (an electrical car could be powered by its own corrosion!). When methane is burnt the change of Gibbs function at 1 atm pressure and 25 °C is -818.0 kJ mol^{-1}, and so 818 kJ of electrical energy can be produced for every 16 g oxidized if it is carried out in a fuel cell, a device for converting the energy of chemical reactions directly into electricity.

The biologically important ATP (adenosine triphosphate) molecule is shown in Fig. 11.9. Its crucial role in metabolism is as a store of energy obtained from food. ATP makes energy available by being able to lose the terminal phosphate group, forming the diphosphate (ADP) and releasing energy in the process. The basic reaction is the hydrolysis of ATP according to

$$\text{ATP} + \text{H}_2\text{O} \rightarrow \text{ADP} + \text{HPO}_4^{2-}(\text{aq}) + 2\text{H}^+(\text{aq});$$

$$\Delta G_{\text{m}}^{\ominus} \approx -30 \text{ kJ mol}^{-1} \quad \text{at } 37\,°\text{C}$$

(37 °C is blood temperature). This indicates that the hydrolysis reaction can provide 30 kJ of useful work for every mole of ATP molecules consumed, which can be used to drive other reactions that are themselves not spontaneous. For instance, the Gibbs function of the reaction in which glucose and fructose are combined to form sucrose has $\Delta G_{\text{m}}^{\ominus} \approx +23$ kJ mol^{-1}. It is therefore not spontaneous. But if the energy of the ATP hydrolysis can be channelled to it, then since 30 kJ are available from 1 mol ATP molecules, the overall change in Gibbs function is -7 kJ mol^{-1} and the overall reaction has a natural tendency to occur. (Whether or not it actually does so depends on the availability of suitable enzymes in the cell.)

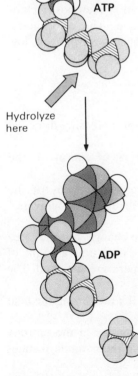

ATP

Hydrolyze here

ADP

11.9 The hydrolysis of ATP to ADP

Some idea of the energy resources required to build proteins can be obtained from the fact that the work from about three ATP hydrolysis reactions is needed to form one peptide link. Even as small a protein as myoglobin contains about 150 peptide linkages, and so every molecule requires for its construction the energy resources of around 450 ATP molecules.

Appendix: The background to entropy

The thermodynamic definition of entropy is in terms of the energy transferred as heat to a sample at a temperature T. The explicit *definition* of an entropy change is

$$\Delta S = q/T, \quad q \text{ transferred reversibly.}$$

ΔS is the change of entropy brought about by a transfer of a quantity of energy q as heat at the temperature T. The term 'reversible' refers to the way the transfer is carried out; it must be carried out extremely carefully, and differences of temperature must not be allowed to develop either between the sample and its surroundings or within the sample itself. We can see that there is a smaller change of entropy when a given quantity of energy is transferred to an object at high temperature than at low: more chaos is introduced when the sample is cool than when it is already hot. The ability of a substance to distribute energy over its particles is related to its heat capacity. This is the basic reason why heat capacity measurements are used to determine entropies. There are other methods, including the use of electrochemical cells and calculation. The calorimetric measurement of entropy involves measuring heat capacities from very low temperatures all the way up to the temperature of interest.

Summary

1 The *natural direction of change* is the direction of increasing dispersal of energy.
2 The *entropy* is a measure of the chaotic dispersal of energy; as the dispersal increases so does the entropy.
3 The *Second law of thermodynamics* states that the entropy of the universe increases in the course of every natural change. The 'universe' means the system and its surroundings.
4 The Second law refers to the natural *tendency* to change, not the rate.
5 The *standard molar entropy*, $S_m^\ominus$, is the entropy of unit amount of substance when it is in its standard state at the temperature specified. The units are normally $J\,K^{-1}\,mol^{-1}$.
6 Entropies can be measured using a calorimeter (by measuring the heat capacity over a range of temperatures).
7 The *standard entropy of reaction*, $\Delta S^\ominus$, is the change in the entropy accompanying complete reaction, all species being in their standard states at the temperature specified.
8 The *change in the entropy of the surroundings* when a reaction occurs with reaction enthalpy ΔH is $\Delta S(\text{surroundings}) = -\Delta H/T$, eqn (11.2.1).

9 A *spontaneous process* is one that may occur without it being necessary to do work. Spontaneous does not mean fast.

10 A process is spontaneous if the *total change of entropy*, the sum of the changes in the system and in the surroundings, is positive.

11 The change in the *Gibbs function*, ΔG, is given by $\Delta G = \Delta H - T\Delta S$ if the temperature is constant, eqn (11.3.2). The Gibbs function, G, is also called the *Gibbs free energy*.

12 A process is spontaneous if it corresponds to a negative value of ΔG.

13 The *standard Gibbs function of reaction*, $\Delta G^{\ominus}$, is the change in the Gibbs function during a complete reaction, all species being present in their standard states at the temperature specified.

14 The *standard molar Gibbs function of formation*, $\Delta G^{\ominus}_{f,m}$, is the change of Gibbs function when unit amount of substance of the compound is formed from its elements, all species being in their standard states at the temperature specified.

15 The standard molar Gibbs function of a reaction is related to the equilibrium constant by $\Delta G^{\ominus}_m = -RT \ln K$, eqn (11.3.3). When $\Delta G^{\ominus}_m$ is large and negative the equilibrium lies strongly in favour of the products; when it is large and positive, the equilibrium lies strongly in favour of the reactants.

16 The change in the Gibbs function is equal to the *maximum quantity of electrical work that can be obtained by harnessing the process*, $\Delta G = w_{e,max}$.

17 One reaction can drive another in its unnatural direction if the former is accompanied by a greater decrease in Gibbs function than the latter, so that the overall ΔG is negative.

Problems

1 Write an account of the natural direction of change. Include as examples the cooling of hot objects, and chemical reactions.

2 Why does a bouncing ball come to rest?

3 State the Second law of thermodynamics (*a*) in terms of entropy, (*b*) without using the word 'entropy'.

4 Why do endothermic reactions occur?

5 Explain, in terms of the changes in the entropy of the system and of the surroundings, why endothermic reactions are favoured by an increase of temperature. (Investigate the effect of the presence of T in the entropy change of the surroundings.)

6 Explain, in terms of entropy and the natural tendency of energy to disperse, why chemical reactions take place spontaneously in the direction corresponding to a decrease in the Gibbs function.

7 Explain why the Gibbs function may also be called the Gibbs free energy.

8 Calculate the standard molar entropy changes at 25 °C for the following reactions: (*a*) $2H_2(g) + O_2(g) \rightarrow 2H_2O(l)$, (*b*) $N_2(g) + 3H_2(g) \rightarrow 2NH_3(g)$, (*c*) $12C(s) + 11H_2O(l) \rightarrow C_{12}H_{22}O_{11}(s)$, sucrose.

9 Calculate the standard molar enthalpy changes at 25 °C for the reactions in the last Question, and then find the values of the standard molar Gibbs functions for the reactions.

10 Calculate the standard molar Gibbs functions of reactions (*a*) and (*b*) in

Question 8 directly from the data in Table 11.2.

11 The standard molar entropy of sucrose is large, but that is partly because each molecule contains so many atoms. What are the entropies of (a) sucrose, (b) copper per unit amount of *atoms* present?

12 The standard molar Gibbs functions of the reactions

$$C + O_2 \rightarrow CO_2; \qquad \Delta G_m^\ominus = -380 \text{ kJ mol}^{-1}$$
$$2C + O_2 \rightarrow 2CO; \qquad \Delta G_m^\ominus = -500 \text{ kJ mol}^{-1}$$

are given for 1500 °C. On the basis of the following information, discuss the possibility of reducing Al_2O_3, FeO, PbO, and CuO with carbon at this temperature.

$$4Al + 3O_2 \rightarrow 2Al_2O_3 \qquad \Delta G_m^\ominus = -2250 \text{ kJ mol}^{-1}$$
$$2Fe + O_2 \rightarrow 2FeO \qquad \Delta G_m^\ominus = -250 \text{ kJ mol}^{-1}$$
$$2Pb + O_2 \rightarrow 2PbO \qquad \Delta G_m^\ominus = -120 \text{ kJ mol}^{-1}$$
$$2Cu + O_2 \rightarrow 2CuO \qquad \Delta G_m^\ominus \approx 0$$

Calculations like this are used in the discussion of metal extraction.

13 Explain why ΔG, the change in the Gibbs free energy, is a better guide to whether a reaction will occur spontaneously than is ΔH, the enthalpy change.

The standard free energies of formation of propene and propane are +62.7 and −23.5 kJ mol^{-1}. Calculate the free energy change for the reaction of propene with molecular hydrogen to give propane, indicate whether the reaction is favourable thermodynamically, and comment on the need for a catalyst when the hydrogenation of propene is carried out in practice.

The addition of an excess of sulphur dichloride oxide (SOCl$_2$, the acid chloride of sulphurous acid) to the hydrated chloride of a transition metal leads to a brisk reaction attended by gas evolution and a colour change, and the temperature of the mixture drops markedly. Comment on the reaction which occurs and on the changes in enthalpy and entropy which accompany it.

(Oxford and Cambridge S)

12

Equilibria involving ions

Now we extend the discussion of the last two chapters to the case of ionic substances in water. First we see how solubilities depend on the conditions, including the presence of other salts. We then turn to a particularly important ion, the hydrogen ion. This ion is responsible for the properties of acids, and is therefore of central importance in chemistry. Once we know about pH we can discuss titrations, hydrolysis, and indicators. We shall also be able to discuss the chemically and biologically important buffer solutions.

Introduction

A lot of chemistry involves the reactions of ions in solution. These reactions include simple precipitation, as when silver chloride is precipitated during analysis, and reactions of considerable economic impact, such as the generation of electricity and the processes involved in corrosion. One ion which is second only to the electron in its importance in chemistry is the hydrogen ion, the proton. Its properties in solution govern the behaviour of acids and bases, substances that play a role in all aspects of daily life, including life itself. Proteins, for instance, can be thought of as being complicated acids and bases and their function is often merely a very elaborate form of titration.

12.1 Dissolving

Ionic compounds dissolve to the point where the solution is *saturated* and no more solid can dissolve. The concentration of the saturated solution is termed the *solubility* of the substance. In some cases the solubility may be very high and a large amount of the solid may dissolve before the solution is saturated. The solubility of common salt, for instance, is 357 g in 1 kg water at 20 °C. In other cases the solubility may be very low. For instance, the solubility of silver chloride is only 7×10^{-6} g in 1 kg water at the same temperature.

The solubility product
The link with the material in the last two chapters is that a saturated solution is in dynamic *equilibrium* with the excess solid present. The

dissolution equilibrium can be expressed in terms of an equilibrium constant. For instance, in the case of silver chloride,

$$AgCl(s) \rightleftharpoons Ag^+(aq) + Cl^-(aq); \qquad K_c = \left\{ \frac{[Ag^+][Cl^-]}{[AgCl]} \right\}_{eq}.$$

As explained in Chapter 10, the concentration of a solid is a constant, and so $[AgCl]$ may be absorbed into K_c. When this is done it is conventional to refer to the resulting quantity as the *solubility product*, and to denote it K_{sp}:

$$K_{sp} = \{[Ag^+][Cl^-]\}_{eq}.$$

Likewise, the solubility product for a salt $M_x A_y$ that dissociates into x cations M^{m+} and y anions A^{a-} is

$$K_{sp} = \{[M^{m+}]^x [A^{a-}]^y\}_{eq}. \tag{12.1.1}$$

Since the solubility product is most useful for the description of the properties of *sparingly soluble salts* (which means salts with solubilities up to about 10^{-3} mol kg^{-1} water) we shall confine the discussion to them.

The solubility products of sparingly soluble salts can be determined in a variety of ways. The aim is to measure the concentrations of the cations and the anions in the saturated solution in contact with excess undissolved solid. The principal techniques involve evaporating a known volume of saturated solution to dryness and then weighing the residue, titration, ion-exchange (which involves replacing cations by protons, Section 10.3, and then titrating), and conductivity. We shall see in the next chapter that an electrochemical cell can also be used.

Example: In an experiment to measure the solubility product of silver chloride the conductivity of a saturated aqueous solution was measured and found to be 1.96×10^{-6} S cm^{-1} at 25 °C. The water itself had a conductivity of 0.12×10^{-6} S cm^{-1} at that temperature. Find the value of K_{sp}.

Method: The aim is to find the concentrations of the Ag^+ and Cl^- ions in the saturated solution; then use $K_{sp} = \{[Ag^+][Cl^-]\}_{eq}$. Find the conductivity due to the ions by taking the difference of the measured values for the solution and the water. Find the molar conductivities from the tables of ion conductivities (Table 8.1). Find the concentration from $c\Lambda_m = \kappa$. Take the molar conductivity as having its value at infinite dilution, since the solution is so dilute.

Answer: The observed conductivity due to the dissolved ions is

$$\kappa = (1.96 \times 10^{-6} \text{ S cm}^{-1}) - (0.12 \times 10^{-6} \text{ S cm}^{-1}) = 1.84 \times 10^{-6} \text{ S cm}^{-1}.$$

The molar conductivity at infinite dilution is

$$\Lambda^\infty = \Lambda^\infty(Ag^+) + \Lambda^\infty(Cl^-) = (61.9 + 76.4) \text{ S cm}^2 \text{ mol}^{-1} = 138.3 \text{ S cm}^2 \text{ mol}^{-1}.$$

Therefore the concentration of ions is

$$c = \frac{1.84 \times 10^{-6}\,\text{S cm}^{-1}}{138.3\,\text{S cm}^2\,\text{mol}^{-1}} = 1.33 \times 10^{-8}\,\text{mol cm}^{-3} = 1.33 \times 10^{-5}\,\text{mol dm}^{-3}$$

(since $10^3\,\text{cm}^3 = 1\,\text{dm}^3$). Since $[Ag^+] = [Cl^-]$ in the electrically neutral solution, we have $[Ag^+] = [Cl^-] = c$. Therefore

$$K_{sp} = (1.33 \times 10^{-5}\,\text{mol dm}^{-3})^2 = 1.77 \times 10^{-10}\,\text{mol}^2\,\text{dm}^{-6}.$$

Comment: Note that the solubility product depends on the temperature, and so in all measurements of K_{sp} the experiment has to be carried out under careful temperature control.

Values of solubility products for some sparingly soluble salts are listed in Table 12.1. A knowledge of K_{sp} lets us make both qualitative and quantitative predictions exactly as in the case of the equilibrium constants of chemical reactions. The point to remember is that at a given temperature the solubility product is a *constant*, independent of the individual ion concentrations. The ion concentrations therefore always tend to adjust so that their product is equal to the value of K_{sp}.

Table 12.1 Solubility products in water at 25 °C

$PbCl_2$	1.6×10^{-5}	$Fe(OH)_2$	7.9×10^{-16}	PbS	1.3×10^{-28}
$BaCO_3$	5.1×10^{-9}	$Zn(OH)_2$	2.0×10^{-17}	$Al(OH)_3$	1.0×10^{-33}
$CaCO_3$	4.8×10^{-9}	AgI	8.3×10^{-17}	$Fe(OH)_3$	2.0×10^{-39}
AgCl	1.8×10^{-10}	FeS	6.3×10^{-18}		
$BaSO_4$	1.3×10^{-10}	CuS	6.3×10^{-36}		

Note that these are the magnitudes of K_{sp} with concentrations expressed in mol dm^{-3}

Experimental measurements of the concentrations of silver and chloride ions in a saturated solution of silver chloride in water at 25 °C give the solubility product at this temperature as

$$K_{sp} = \{[Ag^+][Cl^-]\}_{eq} = 1.77 \times 10^{-10}\,\text{mol}^2\,\text{dm}^{-6}.$$

Consider what happens when more than enough salt needed to form a saturated solution is added. At first there is an imbalance in the rates of dissolution and precipitation. When the concentration has risen sufficiently, a dynamic equilibrium is established and there is no more net dissolution. What is the concentration of salt in solution when this stage is reached? We know the value of the product of the concentrations. The concentrations of chloride and silver ions are equal (because whenever an AgCl unit goes into solution it gives one Ag^+ ion and one Cl^- ion). Therefore $[Ag^+] = [Cl^-]$ at every stage of dissolution, and so at equilibrium

$$K_{sp} = \{[Ag^+][Cl^-]\}_{eq} = [Ag^+]_{eq}^2.$$

The concentration of silver ions in the saturated solution is therefore

$$[Ag^+]_{eq} = \sqrt{K_{sp}} = \sqrt{(1.77 \times 10^{-10} \, mol^2 \, dm^{-6})} = 1.33 \times 10^{-5} \, mol \, dm^{-3}.$$

This is also the concentration of silver chloride in solution at equilibrium, and so we can conclude that the solubility of silver chloride is $1.33 \times 10^{-5} \, mol \, dm^{-3}$. This result also applies to other salts of the same valence type and to other temperatures:

> The solubility of a salt of the type M^+A^- is equal to $\sqrt{K_{sp}}$ at the temperature specified.

The effects of added salts

The solubility product lets us predict what happens when we artificially increase the concentration of one of the ions present in the equilibrium. For instance, what happens when we add sodium chloride to a solution already saturated with silver chloride?

The *qualitative* prediction comes from an application of le Châtelier's principle: increasing the concentration of chloride ions results in the equilibrium shifting so as to minimize the increase. In other words, the sparingly soluble salt precipitates, and we conclude that the sparingly soluble salt is less soluble in the presence of a common ion. This is the *common-ion effect*.

The same conclusion follows from the form of the solubility product, for if we increase $[Cl^-]$ by adding chloride ions, the value of $[Ag^+]$ must decrease in order to preserve the value of K_{sp}, but we can go on to make a *quantitative* prediction about the size of the effect. Suppose we add enough sodium chloride to the solution to make it $1.0 \times 10^{-4} \, mol \, dm^{-3}$ in NaCl (5.8 mg in $1 \, dm^3$ – a few grains). Then $[Cl^-] \approx 1.0 \times 10^{-4} \, mol \, dm^{-3}$, showing that the chloride concentration is dominated by the contribution from the sodium chloride. In order for the solubility product to remain at $1.77 \times 10^{-10} \, mol^2 \, dm^{-6}$ the silver ion concentration must fall to

$$[Ag^+] = \frac{K_{sp}}{[Cl^-]} = \frac{1.77 \times 10^{-10} \, mol^2 \, dm^{-6}}{1.0 \times 10^{-4} \, mol \, dm^{-3}} = 1.77 \times 10^{-6} \, mol \, dm^{-3}.$$

In other words, whereas the solubility of silver chloride in pure water is $1.33 \times 10^{-5} \, mol \, dm^{-3}$, its solubility in $1.0 \times 10^{-4} \, mol \, dm^{-3}$ NaCl solution is only about $1.77 \times 10^{-6} \, mol \, dm^{-3}$, an 8-fold decrease.

The experimentally observed dependence of the solubility on the concentration of added salt is shown in Fig. 12.1. The solubility decreases in line with the prediction we have just made. Yet something strange happens when the added salt concentration exceeds about $0.2 \, mol \, dm^{-3}$; the solubility of silver chloride begins to increase!

The explanation for this behaviour lies in the chemistry of silver. In the presence of excess chloride ions, Ag^+ forms *complex ions* of the form $AgCl_2^-$, $AgCl_3^{2-}$, and $AgCl_4^{3-}$, which are soluble. Not only are they soluble, but as a result of their formation the concentrations of both Ag^+ and Cl^- decrease in the solution; therefore more of the solid goes into solution in order to maintain the solubility product, and so there is an apparent

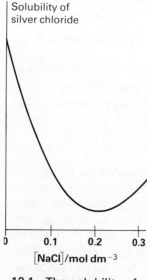

Solubility of silver chloride

$[NaCl]/mol \, dm^{-3}$

12.1 The solubility of silver chloride in the presence of sodium chloride

Principles of physical chemistry

increase in solubility. Similar *complexation reactions* occur with other species. For instance, it accounts for the dissolution of aluminium hydroxide by the addition of excess sodium hydroxide (which complexes the aluminium to form the soluble $Al(OH)_4^-$ ion).

12.2 Acids and bases

The definitions of acids and bases have increased in sophistication since Robert Boyle ascribed the name acid to substances with a sharp taste and interpreted this property in terms of spikes on the atoms.

Brønsted-Lowry acids and bases

One modern definition of acids and bases is due to Brønsted and Lowry. In 1923 they introduced the definitions

> *Brønsted-Lowry acid: a proton donor,*
> *Brønsted-Lowry base: a proton acceptor.*

An *alkali* is a water-soluble base.

Not only do these definitions include 'conventional' acids and bases, but they also go beyond by capturing less obvious substances and labelling them as acid or base. This is typical of physical chemistry: it recognizes that superficially unrelated species may behave in the same way and then generalizes definitions so that they become more widely applicable. In that way the ideas behind the subject become unified and concepts become simplified.

Consider, for example, the nature of ethanoic acid – well, is it an acid? In water ethanoic acid is partially dissociated according to the equilibrium

$$CH_3COOH(aq) + H_2O \rightleftharpoons CH_3COO^-(aq) + H_3O^+(aq).$$

Clearly, the CH_3COOH molecule is a proton donor in water, and so is properly classified as an acid. We can also begin to see how the Brønsted-Lowry definition captures more than just the obvious and the trivial. Note, for instance, that the CH_3COOH donates its proton to an H_2O molecule. That shows that H_2O is a proton acceptor. According to the Brønsted-Lowry definition water is therefore a *base*.

Now turn to the right-hand side of the equilibrium. Since there is an equilibrium (which, like all chemical equilibria, is dynamic) the reverse reaction, the formation of CH_3COOH, occurs when H_3O^+ donates a proton to the CH_3COO^- ion. Therefore, according to the definition of an acid, H_3O^+ is an acid and CH_3COO^- is a base. The ethanoate ion, CH_3COO^-, is called the *conjugate base* of ethanoic acid, CH_3COOH. Similarly, H_3O^+ is called the *conjugate acid* of the base H_2O. Since the ethanoic acid is only weakly dissociated at moderate concentrations, the equilibrium lies in favour of CH_3COOH. A way of expressing this is to regard H_3O^+ as a stronger acid than CH_3COOH, so that it has a stronger tendency to donate a proton. Alternatively, CH_3COO^- can be regarded as a stronger base than H_2O, so that it dominates the accepting of any protons that are available.

Similar remarks apply to species we conventionally regard as bases, such as ammonia. In water the equilibrium established is

$$NH_3(aq) + H_2O \rightleftharpoons NH_4^+(aq) + OH^-(aq).$$

NH_3 is clearly a proton acceptor, and is therefore a base. In this reaction, however, the H_2O is the donor of the proton, not the acceptor. Therefore it is acting as an *acid*. Note at this stage how water can act both as an acid and as a base; that is a centrally important feature. The reverse reaction in the equilibrium is the donation of a proton by the ammonium ion, NH_4^+, to a hydroxide ion, OH^-. It follows that NH_4^+ is an acid, the conjugate acid of the base NH_3, while OH^- is the conjugate base of the acid H_2O. Since the equilibrium lies predominantly to the left, OH^- is a stronger base than NH_3; or, alternatively, NH_4^+ is a stronger acid than H_2O.

We have emphasized that water can act either as an acid or as a base, depending on the demands of the other species present. It is not alone in this property. Consider, for instance, what happens when hydrogen chloride dissolves in pure ethanoic acid. The following equilibrium is established:

$$CH_3COOH + HCl \rightleftharpoons CH_3COOH_2^+ + Cl^-.$$

HCl is a strong proton donor (which is why it behaves as a strong acid in water); CH_3COOH is a relatively weak donor. Therefore, in this system, HCl acts as the donor while CH_3COOH acts as the acceptor; that is, 'ethanoic acid' is a base! The important conclusion to draw is that *whether a substance is an acid or a base depends on the other species present*. A lot of modern chemistry is concerned with *non-aqueous solvents*, and a correct understanding of the roles of species as acids or as bases is essential.

Finally, consider the case of water alone. In the pure liquid there is the equilibrium

$$H_2O + H_2O \rightleftharpoons H_3O^+(aq) + OH^-(aq).$$

Here we see H_2O acting simultaneously as an acid (in donating H^+ leaving OH^-, its conjugate base) and as a base (in accepting H^+ to form H_3O^+, its conjugate acid). In pure water the concentrations of hydrogen ions and hydroxide ions are equal and the liquid is *neutral*. If a proton donor is present $[H_3O^+]$ exceeds $[OH^-]$, and the solution is *acidic*. When a proton acceptor is present $[OH^-]$ exceeds $[H_3O^+]$, and the solution is *basic*.

Dissociation constants

All the equations we have written in this section are *equilibria*. Therefore we can use the techniques developed in Chapter 10 to describe them; we introduce equilibrium constants, and then explore their implications.

Consider the equilibrium established in aqueous ethanoic acid:

$$CH_3COOH(aq) + H_2O \rightleftharpoons CH_3COO^-(aq) + H_3O^+(aq);$$

$$K_c = \left\{ \frac{[H_3O^+][CH_3COO^-]}{[H_2O][CH_3COOH]} \right\}_{eq}.$$

The concentration of water is almost constant on account of the relatively small amount involved in the dissociation. Therefore we absorb $[H_2O]$ into the equilibrium constant and write the new constant as the *acid dissociation constant*, K_a:

$$K_a = \left\{ \frac{[H_3O^+][CH_3COO^-]}{[CH_3COOH]} \right\}_{eq}. \qquad (12.2.1)$$

The value of K_a for ethanoic acid is only 1.7×10^{-5} mol dm^{-3}, and the small extent of dissociation this represents means that it is a 'weak' acid. Some species can donate more than one proton and are called *polyprotic acids*. For instance, H_2SO_4 can donate up to two protons, and hence is a *diprotic acid* (or, since it has two conjugate bases, HSO_4^- and SO_4^{2-}, a *dibasic acid*). In solution there are several equilibria between the acid and its conjugate bases, and they are described by the *first, second, etc., dissociation constants*. The second dissociation constant is usually much smaller than the first because it is energetically more difficult to remove a second proton from a negatively charged species (*e.g.*, from HSO_4^-). A major application of the study of polyprotic acids is to the function of proteins, for they are naturally occurring polyprotic species.

The most important characteristic of an aqueous solution of an acid is the hydrogen ion concentration $[H_3O^+]$. An approximate way of relating $[H_3O^+]$ to K_a is to suppose that a weak acid dissociates so little that the concentration of the undissociated form (*e.g.*, CH_3COOH) is equal to the concentration originally added. We shall write $[CH_3COOH]_{eq} \approx [CH_3COOH]_{added} = A$. Moreover, if we ignore the relatively slight dissociation of water, the hydrogen ion and ethanoate ion concentrations are equal, because one of each type of ion is formed whenever an acid molecule dissociates. We therefore write $[CH_3COO^-]_{eq} \approx [H_3O^+]_{eq}$. These two approximations turn the last equation into

$$K_a \approx \frac{[H_3O^+]_{eq}^2}{A}; \quad \text{or} \quad [H_3O^+]_{eq} \approx \sqrt{(K_a A)}. \qquad (12.2.2)$$

If we have an independent source of information about the value of K_a,

Table 12.2 Acid dissociation constants in water at 25 °C, $K_a/$(mol dm^{-3}) and pK_a

	$K_a/$mol dm^{-3}	pK_a
CH_3COOH	1.74×10^{-5}	4.76
HF	5.62×10^{-4}	3.25
HOCl	3.72×10^{-8}	7.43
H_2SO_4	1.20×10^{-2}	1.92 (pK_1)
H_3PO_4	7.08×10^{-3}	2.15 (pK_1)
	6.17×10^{-8}	7.21 (pK_2)
	4.37×10^{-13}	12.36 (pK_3)

or access to tables of data (such as those in Table 12.2), it is easy to predict the hydrogen ion concentration in a solution of known acid concentration.

Acids and pH

Even in quite ordinary chemistry the value of $[H_3O^+]$ can vary over many orders of magnitude. As a consequence it is common to discuss not the concentration itself but its *logarithm*. It turns out to be sensible to introduce a quantity called the *pH* of a solution (it was originally introduced by Sørensen in an attempt to improve the quality control of beer). pH is defined as the *negative* of the logarithm of the hydrogen ion concentration:

$$pH \text{ of a solution: } pH = -\lg [H_3O^+]. \tag{12.2.3}$$

The log is to the base 10 ('ordinary' logarithms). Since it is impossible to take the log of a quantity with dimensions we must interpret $[H_3O^+]$ in this expression as the numerical value of the concentration when it is expressed in $mol\,dm^{-3}$. A better definition, which is explicit but clumsy to keep writing, is therefore

$$pH = -\lg \{[H_3O^+]/mol\,dm^{-3}\}. \tag{12.2.4}$$

We shall use this form in the *Examples* because it eliminates any worries about units.

Consider a solution of ethanoic acid which is so dilute that we can use the simple expression in eqn (12.2.2). The pH is given by

$$pH = -\lg [H_3O^+] = -\lg \sqrt{(K_a A)} = -\tfrac{1}{2}\lg K_a A = -\tfrac{1}{2}\lg K_a - \tfrac{1}{2}\lg A.$$

(Remember that $\lg \sqrt{x} = \tfrac{1}{2}\lg x$ and $\lg ab = \lg a + \lg b$.) The term $-\lg K_a$ resembles the expression defining pH and from now on it will be written pK_a:

$$pK_a = -\lg K_a. \tag{12.2.5}$$

(Not only is this tidy, which is always an advantage for reading complicated expressions, but it has the effect of removing factors like 10^{-5} from the tables of dissociation constants, Table 12.2.) Then

$$pH = \tfrac{1}{2}pK_a - \tfrac{1}{2}\lg A. \tag{12.2.6}$$

Example: Calculate the pH of the following aqueous solutions: (*a*) 2.0 mol dm^{-3} HCl(aq), (*b*) 0.1 mol dm^{-3} HCl(aq), (*c*) 0.1 mol dm^{-3} CH_3COOH(aq) at 25 °C.

Method: Base the answer on eqn (12.2.4) for the fully dissociated acid (HCl) and on eqn (12.2.6) for the partially dissociated acid (CH_3COOH). For (*c*) it is necessary to know the value of pK_a: use Table 12.2.

Answer: Since hydrogen chloride is almost completely dissociated in aqueous solution we use

---

Principles of physical chemistry

Alkalis and pH

The next step is to apply the same arguments to the dissociation equilibria involving alkalis, and in particular to weak (that is, slightly dissociated) bases such as ammonia in water. The equilibrium when ammonia is dissolved in water is

$$NH_3(aq) + H_2O \rightleftharpoons NH_4^+(aq) + OH^-(aq); \qquad K_c = \left\{\frac{[NH_4^+][OH^-]}{[NH_3][H_2O]}\right\}_{eq}.$$

As in the case of acid dissociations it is sensible to absorb the water concentration into the constant, and therefore to define the *base dissociation constant* and its logarithm:

$$K_b = \left\{\frac{[NH_4^+][OH^-]}{[NH_3]}\right\}_{eq}; \qquad pK_b = -\lg K_b. \qquad (12.2.7)$$

Some values are listed in Table 12.3.

Table 12.3 Base dissociation constants in water at $25\,°C$, $K_b/(\text{mol dm}^{-3})$ and pK_b

	$K_b/\text{mol dm}^{-3}$	pK_b
NH_3	1.74×10^{-5}	4.76
$C_2H_5 \cdot NH_2$	5.62×10^{-4}	3.25
$(C_2H_5)_2 \cdot NH$	9.55×10^{-4}	3.02
$C_6H_5 \cdot NH_2$	3.80×10^{-10}	9.42

The OH^- ion concentration can be calculated from the value of K_b and the concentration of base added initially, but it turns out to be much more useful to express the properties of aqueous solutions of bases in terms of the *hydrogen ion* concentration. This may seem curious, for where do we refer to hydrogen ions in the base dissociation equilibrium?

The problem is resolved once it is realized that the base is present in water, and that water itself dissociates slightly to produce hydrogen ions:

$$H_2O + H_2O \rightleftharpoons H_3O^+(aq) + OH^-(aq); \qquad K_c = \left\{\frac{[H_3O^+][OH^-]}{[H_2O]^2}\right\}_{eq}.$$

$$(12.2.8)$$

As in the case of weak acids and bases, the dissociation is so slight that $[H_2O]$ is approximately constant, and can be absorbed into the equilibrium constant. This gives the *ionic product of water*, K_w, and its logarithm:

Ionic product of water: $K_w = \{[H_3O^+][OH^-]\}_{eq}; \qquad pK_w = -\lg K_w.$

$$(12.2.9)$$

At $25\,°C$ $K_w = 1.0 \times 10^{-14}\,mol^2\,dm^{-6}$ ($pK_w = 14.0$) reflecting water's very low degree of dissociation.

The water dissociation equilibrium has several consequences. In the first place we can immediately state the hydrogen ion concentration in pure water from a knowledge of the value of K_w. Since both an OH^- and an H_3O^+ ion are formed whenever an H_2O molecule dissociates, in the neutral solution $[H_3O^+] = [OH^-]$, and so

$$[H_3O^+] = \sqrt{K_w} = 1.0 \times 10^{-7}\,mol\,dm^{-3} \quad \text{at } 25\,°C.$$

It follows that the pH of pure water at $25\,°C$ is

$$pH = -\lg[H_3O^+] = -\lg(1.0 \times 10^{-7}) = 7.0.$$

In other words, *the pH of a neutral solution at 25 °C is 7.0.* This is an extremely important point, and will recur throughout the remainder of this discussion.

Since K_w is a constant, when $[OH^-]$ increases in the presence of a base, $[H_3O^+]$ must decrease, and vice versa. In particular the two concentrations are related by $[H_3O^+] = K_w/[OH^-]$. Therefore equilibria involving alkalis can be expressed in terms of the hydrogen ion concentration. This is a very important chain of argument, and we shall now show what it means in practice.

When a base dissolves in water the dissociation equilibrium is established and OH^- ions are generated. In the case of ammonia, for instance, since the dissociation is so small, the value of $[NH_3]$ in the expression for K_b is almost equal to the concentration of ammonia added ($[NH_3]_{eq} \approx [NH_3]_{added} = B$) and the concentrations of OH^- and NH_4^+ are approximately the same; therefore eqn (12.2.7) gives

$$K_b \approx \frac{[OH^-]_{eq}^2}{B}, \quad \text{or} \quad [OH^-]_{eq} \approx \sqrt{(K_b B)}.$$

The water dissociation equilibrium adjusts so as to maintain $[H_3O^+] = K_w/[OH^-]$. Then following the same argument as before, we find

$$pH = pK_w - \tfrac{1}{2}pK_b + \tfrac{1}{2}\lg B \qquad (12.2.10)$$

and we can calculate the pH of any weak alkali.

Principles of physical chemistry

Example: Calculate the pH of (*a*) 0.1 mol dm^{-3} NH$_3$(aq) at 25 °C, (*b*) the same concentration of NaOH(aq).

Method: For the partially dissociated ammonia, use eqn (12.2.10) with pK_b taken from Table 12.3. The pH of a strong base can be obtained on the basis that it is completely dissociated, and that [OH$^-$] = [NaOH]; then express [OH$^-$] in terms of [H$_3$O$^+$] through K_w/[OH$^-$] = [H$_3$O$^+$].

Answer: (*a*) Since pK_b = 4.76, pK_w = 14.0, and B = 0.1 mol dm^{-3} use of eqn (12.2.10) gives

$$\text{pH} = \text{p}K_w - \tfrac{1}{2}\text{p}K_b + \tfrac{1}{2}\lg B = 14.0 - \tfrac{1}{2} \times 4.76 + \tfrac{1}{2}\lg 0.1 = 11.1.$$

(*b*) In the case of the strong alkali at the same concentration,

$$[\text{H}_3\text{O}^+] = K_w/[\text{OH}^-], \quad \text{so that} \quad \text{pH} = \text{p}K_w + \lg [\text{OH}^-] = \text{p}K_w + \lg B,$$

where B is the concentration of base added. In the present case,

$$\text{pH} = 14.0 + \lg 0.1 = 13.0.$$

Comment: Both aqueous ammonia and aqueous sodium hydroxide are key industrial and laboratory alkalis. Note how the pH of the strong base is higher than that for ammonia even though the nominal concentrations are the same; this is because ammonia is only partially dissociated, the OH$^-$ concentration is lower, and therefore (on account of the water dissociation equilibrium) the hydrogen ion concentration is higher.

12.3 Applications of pH

The pH of a solution is simply another way of expressing the hydrogen ion concentration. From its definition, and as we have seen in the *Examples*, at 25 °C:

the pH of an acidic solution is less than 7.0,
the pH of a neutral solution is 7.0,
the pH of an alkaline solution is greater than 7.0.

There are no upper or lower limits to the values of pH, but for most applications it has values in the range 0 to 14, Fig. 12.2, and so most devices for measuring pH (which we shall meet in the next Chapter) are calibrated in that range.

Hydrolysis

We can now account for the behaviour of salts of weak acids and strong bases, and of strong acids and weak bases, when they are dissolved in water. All we have to bear in mind is that the hydrogen ion concentration adjusts so as to maintain the constancy *both* of the acid or base dissociation constant *and* of the ionic product of water.

Consider the case of sodium ethanoate in water. This is the salt of a weak acid (ethanoic acid, pK_a = 4.76) and a strong base (sodium hydrox-

pH

14 —

12 —

10 —

8 —

7 — ◄— Neutral

6 —

4 —

2 —

0 —

Basic

Acid

Strongly acid

12.2 The pH scale

ide; fully dissociated). When it dissolves it gives rise to ethanoate ions, CH_3COO^-. In aqueous solution there are two important equilibria:

$$H_2O + H_2O \rightleftharpoons H_3O^+(aq) + OH^-(aq); \qquad K_w = \{[H_3O^+][OH^-]\}_{eq}.$$

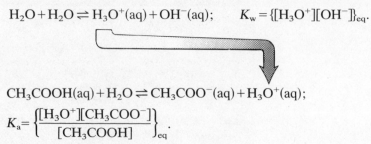

$$CH_3COOH(aq) + H_2O \rightleftharpoons CH_3COO^-(aq) + H_3O^+(aq);$$

$$K_a = \left\{ \frac{[H_3O^+][CH_3COO^-]}{[CH_3COOH]} \right\}_{eq}.$$

The second equilibrium lies strongly in favour of the ethanoic acid. Therefore, when ethanoate ions are first added, the reaction shifts to the left, using hydrogen ions provided by the first equilibrium. This takes place until enough CH_3COOH has been formed and the value of K_a is reached. In order to keep K_w constant, the first equilibrium has to shift to the right. Since the H_3O^+ ions are consumed, an excess of OH^- ions is left in the solution. As a result of this hydrolysis, the solution of the salt is slightly basic, and its pH is greater than 7.0. This is what is observed in practice, and a $0.1 \, mol \, dm^{-3}$ solution of sodium ethanoate is observed to have pH = 7.8.

A solution of a salt of a strong acid and a weak base, such as ammonium chloride, can be expected to show the opposite behaviour, and its hydrolysis can be expected to lead to a slightly acid solution. As a result of the presence of ammonium ions in the solution, the system establishes the hydrolysis equilibrium:

$$H_2O + H_2O \rightleftharpoons H_3O^+(aq) + OH^-(aq)$$

$$NH_3(aq) + H_2O \rightleftharpoons NH_4^+(aq) + OH^-(aq);$$
$$K_b = \{[NH_4^+][OH^-]/[NH_3]\}_{eq}.$$

Since ammonia is a weak base ($pK_b = 4.76$) the equilibrium lies strongly in favour of the left, and so the OH^- ions provided by the water dissociation are removed. The hydrogen ion concentration rises in order to keep K_w constant, and as a result the pH of the solution falls below 7.0. A solution of a salt of a weak base and a strong acid is therefore slightly acidic. This is observed in practice, and a $0.1 \, mol \, dm^{-3}$ solution of ammonium chloride in water has pH = 6.0 at 25 °C.

Titrations

The importance of hydrolysis equilibria lies in their applications, for example to the *end-point of titrations*. At the end-point of an acid-base titration, exactly enough acid has been added to turn all the alkali initially present into a salt. In other words, at the exact end-point there is a solution of a salt.

In the case of a strong acid/strong base titration, the salt produced is

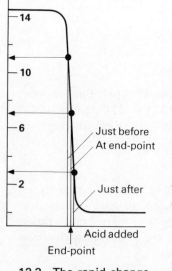

12.3 The rapid change of pH at the end-point

Principles of physical chemistry

not hydrolysed and the only significant equilibrium in solution is the dissociation of water. The solution is neutral, and its pH is 7.0. Hence, if we can measure the pH of the solution in the course of the titration, we can determine the end-point by noting when pH = 7.0.

The detection of the end-point is helped by the fact that pH changes very rapidly there, Fig. 12.3. Suppose the titration is a few drops away from the end-point, and the remaining alkali concentration is only $0.001 \, mol \, dm^{-3}$; its pH is then 11.0 (from $[H_3O^+] = K_w/[OH^-]$). At the end-point it is 7.0. Just after the end-point, when acid is in slight excess, of the order of $0.001 \, mol \, dm^{-3}$ for instance, the pH is about 3.0. Therefore, that very small shift of composition drives the pH rapidly from pH = 11.0 down to 3.0; hence the sharp fall of pH shown in Fig. 12.3 and the ease with which the end-point can be detected.

In the case of a weak base/strong acid titration, the end-point also corresponds to a solution of the salt, but as a result of hydrolysis its pH lies on the acid side of neutrality. The end-point is therefore at some value of pH less than 7.0, as indicated in Fig. 12.4(a). The actual position can be detected by noting where the pH changes rapidly. Similar remarks apply to a strong base/weak acid titration, the difference being that the end-point lies to higher, more basic, values of pH, as indicated in Fig. 12.4(b).

The case of the titration of a weak acid/weak base is shown in Fig. 12.4(c). This is more complicated to deal with because there are now three equilibria involved at the end-point, and the sluggish variation of the pH makes the end-point difficult to detect (and conductimetry, Section 8.3, is used instead).

Buffers
We have seen that at the end-point there is a rapid change of pH for even small changes of concentration. In contrast, in the regions where there are

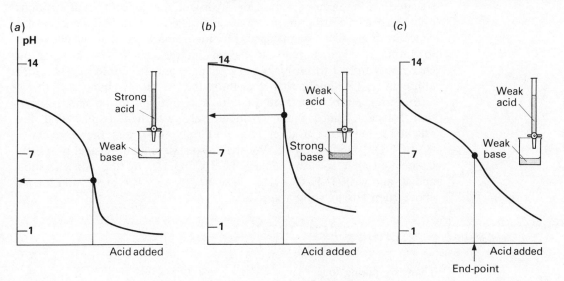

12.4 Potentiometric titrations

approximately equal concentrations of acid and salt, the pH is very insensitive to the acid added. This *buffer region* is indicated in Fig. 12.5.

Buffer solutions are of immense importance in living systems because most metabolic processes have to occur at strictly controlled pH. A part of the reason is that the function of proteins depends on their structures, and structures depend on pH. This is because proteins are both acids and bases, and can donate or accept protons from the medium around them. When they do so they acquire electric charges, and electrostatic interactions then force the structure into a new shape. Only if that shape is correct will the metabolic process occur, otherwise the organism may die. For instance, blood has to be maintained in the range pH = 7.0 to 7.9, saliva functions at pH = 6.8, and the enzymes acting in the stomach require an acid environment in the region of pH = 1.6 to 1.8. All these conditions are buffered in the way we shall now describe.

Suppose a solution consists of approximately equal concentrations of a weak acid (ethanoic acid, for instance) and its salt (such as sodium ethanoate). There is an abundant supply both of ethanoate ions (from the salt) and of the undissociated acid. The equilibrium

$$CH_3COOH(aq) + H_2O \rightleftharpoons CH_3COO^-(aq) + H_3O^+(aq);$$

$$K_a = \{[CH_3COO^-][H_3O^+]/[CH_3COOH]\}_{eq}$$

is established. When a little more acid is added to this solution, the hydrogen ion concentration increase is opposed (by le Châtelier's principle, the equilibrium attempts to minimize the change). This is possible because there are many ethanoate ions and the equilibrium can shift strongly to the left. The change in H_3O^+ concentration is therefore not as marked as when acid is added to water alone. Likewise, if a base is added, the equilibrium responds by shifting to the right and opposing the loss of the hydrogen ions. This is possible because there are plenty of undissociated acid molecules. Hence the hydrogen ion concentration responds only sluggishly. Furthermore, we are interested in pH, the *logarithm* of a concentration. The significance of this remark is that whenever a logarithm is taken, it tends to flatten out the change (for instance on changing x from 1 to 1 000 000, $\lg x$ changes only from 0 to 6). Therefore although $[H_3O^+]$ might change slightly, the pH barely changes at all.

Buffering can be expressed quantitatively in terms of the acid dissociation constant. Since there is added salt, we shall assume that the ethanoate ion concentration is due entirely to it, and therefore write $[CH_3COO^-] \approx [Salt]_{added} = S$. We shall also assume that the acid is so little dissociated that we can replace $[CH_3COOH]_{eq}$ by the concentration added, and write $[CH_3COOH]_{eq} \approx [Acid]_{added} = A$. The expression for K_a above then lets us write

$$[H_3O^+] = K_a \left\{ \frac{[CH_3COOH]}{[CH_3COO^-]} \right\}_{eq} \approx K_a A/S.$$

Then, by taking logs,

$$pH = pK_a - \lg(A/S). \tag{12.3.1}$$

pH

14

7

buffer pH

1

Acid added

Salt   Salt and acid

12.5   The buffer region

Since the logarithm is zero when $A = S$, measurement of the pH of a solution containing equal concentrations of acid and salt gives the pK_a of the acid.

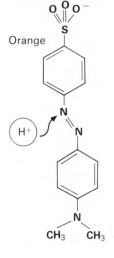

Orange

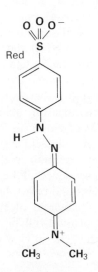

Red

12.6 Methyl orange in its base (orange) and conjugate acid (red) forms

Indicators

A common way of detecting the end-point of an acid–base titration is to use an *indicator*, a substance (a dye) that changes colour according to the pH of the medium. The colour change arises from the attachment or removal of a proton from an important part of the molecule, Fig. 12.6. We shall denote the protonated form as HIn and the unprotonated form as In^-. Then HIn, a weak acid, is in equilibrium with its conjugate base In^-:

$$HIn(aq) + H_2O \rightleftharpoons H_3O^+(aq) + In^-(aq); \qquad K_{In} = \left\{ \frac{[H_3O^+][In^-]}{[HIn]} \right\}_{eq}.$$

As usual, we have absorbed the water concentration into the equilibrium constant and introduced the *indicator dissociation constant, K_{In}*.

The interpretation of how an indicator responds to pH follows the usual arguments about equilibria. For instance, suppose the pH is in the strongly acid region (pH near 1), then there are many hydrogen ions in the solution. Therefore, in order to maintain the value of K_{In}, $[In^-]$ must fall and $[HIn]$ must rise. Consequently, in acid solution the indicator is mainly in the protonated form and shows its characteristic 'acid' colour. In contrast, in basic solution, when the concentration of hydrogen ions is low, the indicator equilibrium requires [In⁻] to increase and [HIn] to decrease. Therefore the indicator is then mainly present as In⁻, and shows its characteristic 'basic' colour. It follows that when the solution swings from low pH to high (or vice versa) the indicator swings from being predominantly HIn to In⁻ (or vice versa), and we see the corresponding change of colour.

The precise range over which an indicator changes predominantly from one form to another depends on its structure, Fig. 12.7. This is valuable when we see it in relation to the discussion about the end-points of different types of acid–base titration. For instance, we have seen that as a result of hydrolysis the titration of a weak base and a strong acid has an end-point in the acid region. Therefore an indicator is needed that changes colour at around pH = 4. Figure 12.7 shows that methyl orange is suitable. In contrast, for the basic end-point of a weak acid/strong base titration, an indicator with a colour change around pH = 9 must be chosen; phenolphthalein is suitable.

12.4 Lewis acids and bases

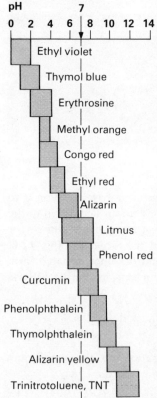

12.7 The pH range for the colour changes of various indicators

We have concentrated on the Brønsted-Lowry definition of acids and bases. This, it should be recalled, is a generalization of a primitive definition of acids and bases. There is, however, an even more general definition. It captures everything we have done so far but is much broader in scope. Breadth of scope, as we have stressed before, means simplification because it unifies what may seem to be quite different concepts.

The *Lewis definition* (introduced by G. N. Lewis in 1923) is that *an acid is an electron pair acceptor*. In contrast, *a base is an electron pair donor*. We can see that this includes Brønsted-Lowry acids and bases, because protons can stick to lone pairs of electrons (think of NH_3) and so an electron pair donor is also a proton acceptor. Likewise, the outstanding example of an electron pair acceptor is a proton, because it can attach to any electron pair that happens to be available. Therefore the proton acts as a Lewis acid, and species that can provide protons constitute acids, as in the Brønsted-Lowry theory (but with a slight shift of emphasis).

The important point to note is that the Lewis definition of an acid and a base does not refer explicitly to the proton. It switches attention to the electronic structure of the compounds. This has two consequences. One is that even species without protons can be classified as acids and bases. For instance, Cl^- is a base; Fe^{3+} is an acid. Therefore the whole of the chemistry of acids and bases can be regarded in terms of the availability of *electron pairs* and this fits much more naturally into the general view of chemistry that it is concerned with the behaviour of electrons.

Summary

1 A *saturated solution* is one in which no more of the solute will dissolve; the solution is *at equilibrium* with undissolved solid.

2   The *solubility product* for a salt M_xA_y is $K_{sp} = \{[M^{m+}]^x[A^{a-}]^y\}_{eq}$. Solubility products are important for *sparingly soluble salts*, and can be found by measuring the ion concentrations in the saturated solutions.

3   The *solubility* of a sparingly soluble salt M^+A^- is equal to $\sqrt{K_{sp}}$ at the temperature specified.

4   Addition of a *common ion* causes precipitation of a sparingly soluble salt. Increased solubility when a common ion is added indicates the formation of a *complex ion*.

5   A *Brønsted-Lowry acid* is a proton donor. A *Brønsted-Lowry base* is a proton acceptor. An *alkali* is a water-soluble base.

6   To every acid there is a *conjugate base*, formed by loss of a proton. To every base there is a *conjugate acid* formed by gain of a proton.

7   The classification of a species as an acid or as a base depends on the other species present in the solution.

8   The *acid dissociation constant* of an acid HA is $K_a = \{[H_3O^+][A^-]/[HA]\}_{eq}$; by definition, $pK_a = -\lg K_a$.

9   The *pH* of a solution is defined as $pH = -\lg\{[H_3O^+]/mol\,dm^{-3}\}$. The pH of neutral water is 7.0 (at 25 °C). Acid solutions have pH less than 7.0 while alkaline solutions have pH greater than 7.0.

10   The *base dissociation constant* of a base B is $K_b = \{[BH^+][OH^-]/[B]\}_{eq}$; by definition, $pK_b = -\lg K_b$.

11   The *ionic product* of water is $K_w = \{[H_3O^+][OH^-]\}_{eq}$; the concentrations of hydrogen and hydroxide ions are linked by this equilibrium.

12   The salts of weak acids/strong bases are *hydrolysed* in water and are mildly basic. The salts of weak bases/strong acids are also hydrolysed, and are mildly acidic.

13   The end-point of an *acid–base titration*, except for weak acids/weak alkalis, can be determined using pH measurement; the pH changes sharply at the end-point.

14   A buffer solution maintains essentially constant pH; an acid buffer consists of a solution of a weak acid and one of its salts in approximately equal concentrations.

15   Titration end-points are frequently determined using *indicators*, dyes with colours that depend on pH. The selection of the indicator must take into account the effects of hydrolysis.

16   *Polyprotic acids* can donate more than one proton.

17   A *Lewis acid* is an electron pair acceptor; a *Lewis base* is an electron pair donor. This represents a generalization of acid–base classification.

Problems

1   Describe what happens when common salt is added to a saturated solution of silver chloride in water.

2   Explain the following observations, stating the natures and formulas of any complex ions involved: (*a*) hydrated copper(II) sulphate is pale blue, (*b*) when excess aqueous ammonia is added to an aqueous solution of copper(II) ions the solution becomes deep blue, (*c*) when concentrated hydrochloric acid is added to a solution of copper(II) ions, the colour changes to green.

3   Discuss the effect of temperature on the solubility of a sparingly soluble salt in the same terms as were used for the effect of temperature on an equilibrium, Chapter 10. The enthalpy of dissolution of silver chloride is $+65\,kJ\,mol^{-1}$. Is silver chloride more or less soluble at higher temperatures?

4   The solubility product of zinc hydroxide is given in Table 12.1. What is its solubility? What is its solubility in $1.0 \times 10^{-4}\,mol\,dm^{-3}$ sodium hydroxide solution?

5   The conductivity of a $0.0634\,mol\,dm^{-3}$ solution of 2-hydroxypropanoic acid (the acid responsible for 'cramp') was measured as $1.138 \times 10^{-3}\,S\,cm^{-1}$. Its molar conductivity at infinite dilution is $388.5\,S\,cm^2\,mol^{-1}$. What is (a) the pH of the solution, (b) the acid dissociation constant?

6   Calculate the pH of $0.1\,mol\,dm^{-3}$ sulphuric acid.

7   Explain why pK_w increases with temperature.

8   Define the terms *Brønsted-Lowry acid* and *Brønsted-Lowry base*. Classify as acids or bases, where possible, the following species: (a) hydrogen chloride in water, (b) sodamide ($NaNH_2$) in liquid ammonia, (c) iron(III) chloride in water.

9   Outline the Lewis definition of acids and bases, and classify, where possible, the three substances in the last question, giving reasons.

10   At 20 °C, the solubility product of strontium sulphate is $4.0 \times 10^{-7}\,mol^2\,dm^{-6}$, and that of magnesium fluoride is $7.2 \times 10^{-9}\,mol^3\,dm^{-9}$. Estimate to two significant figures the solubility at 20 °C in $mol\,dm^{-3}$ of

   (a) strontium sulphate in a 0.1 M solution of sodium sulphate,

   (b) magnesium fluoride in a 0.2 M solution of sodium fluoride.

(Oxford)

11   Distinguish carefully between the terms *solubility* and *solubility product* as applied to a sparingly soluble electrolyte.

What happens when aqueous solutions of the following substances are added to a saturated aqueous solution of lead chloride containing a small excess of undissolved solid?

   (a) sodium chloride,

   (b) sodium iodide,

   (c) silver nitrate.

Account quantitatively for the fact that copper(II) sulphide can be precipitated by hydrogen sulphide from an acidic solution (assume $0.1\,mol\,dm^{-3}$ of free H^+) whereas iron(II) sulphide cannot be precipitated under these conditions.

(Oxford and Cambridge S part question)

12   (a) In each of the following reactions one of the reactants is behaving as a base. In each case underline the species you believe to be the base.

   (i) $HSO_4^- + HNO_2 \rightarrow H_2NO_2^+ + SO_4^{2-}$

   (ii) $H_2PO_4^- + HCO_3^- \rightarrow HPO_4^{2-} + H_2O + CO_2$

   (iii) $CH_3CO_2H + HNO_3 \rightarrow CH_3CO_2H_2^+ + NO_3^-$

   (iv) $HBr + HCl \rightarrow Br^- + H_2Cl^+$

   (b)   (i) Using any one of the systems given above as an example, give and explain briefly the Brønsted-Lowry theory of acids and bases.

      (ii) According to the theory, water is able to function both as an acid and as a base. Give ONE reaction in which water functions as an acid and ANOTHER in which it functions as a base.

   (c) Comment on the fact that a mixture of pure nitric acid, HNO_3, and pure perchloric acid, $HClO_4$, contains the ions $H_2NO_3^+$ and ClO_4^-.

(London)

13 (a) Define the terms *Brønsted acid* and *Brønsted base*.

(b) Explain whether each of the following species normally acts as an acid or a base or both in (i) aqueous solution and (ii) liquid ammonia: NH_4^+, HSO_4^-, NH_2^-, NH_3.

(c) Discuss the factors which have to be taken into account in determining the equivalence point (end-point) in the titration of an aqueous solution of ethanoic acid (acetic acid) with sodium hydroxide solution.

(*Welsh*)

14 (a) Explain why an aqueous solution of ammonium chloride is acidic.

(b) When aqueous solutions of ammonium chloride and ammonia are mixed, a buffer solution can be produced.

(i) What is a buffer solution?

(ii) Explain the buffer action of this mixture.

(c) What is the hydrogen ion concentration of a 0.01 M aqueous solution of methylamine at 25 °C? K_b for methylamine is 4×10^{-4} mol l^{-1} [mol dm^{-3}] and K_w for water is 10^{-14} mol^2 l^{-2} [mol^2 dm^{-6}] both at 25 °C.

(*JMB*)

15 What is the definition of pH? Describe how you would measure the pH of an unknown solution (*not* using an indicator).

Discuss and evaluate the following statements.

(a) More hydrogen will be evolved when excess zinc is added to 50 cm^3 of hydrochloric acid of concentration 0.100 mol dm^{-3} than when excess zinc is added to an equal volume of ethanoic (acetic) acid of the same concentration.

(b) The same volume of a standard solution of sodium hydroxide is needed to neutralize equal volumes of hydrochloric acid and ethanoic acid of the same concentration.

Calculate the pH of the following solutions:

(i) nitric acid of concentration 0.025 mol dm^{-3},

(ii) a saturated solution of 4-hydroxybenzoic acid which contains 6.50 g dm^{-3} at 25 °C (p$K_a = 4.58$).

(iii) a buffer solution made by mixing 60 cm^3 of ethanoic acid of concentration 0.100 mol dm^{-3} with 40 cm^3 of sodium hydroxide of the same concentration.

(*Oxford and Cambridge*)

16 Explain qualitatively why aqueous solutions of sodium ethanoate are not neutral. A hydrolysis constant K_h can be defined as shown below, and has the value given at 298 K.

$$K_h = \frac{[CH_3CO_2H][OH^-]}{[CH_3CO_2^-]} = 5.7 \times 10^{-10} \text{ mol dm}^{-3}.$$

Calculate the concentration of hydroxide ion in an 0.08 M [mol dm^{-3}] aqueous solution of sodium ethanoate at 298 K.

Some information about three titration indicators is shown in the Table.

Table

Name	Colour in 1 M acid	Colour in 1 M alkali	Useful pH range
Bromocresol green	yellow	blue	3.8–5.4
Phenolphthalein	colourless	pink	8.3–10.0
4-Nitrophenol	colourless	yellow	5.6–7.6

(*i*) Explain why 4-nitrophenol can be used as an indicator for acid–base titrations.

(*ii*) Which of the indicators in the Table would be the most appropriate to use in the titration of ethanoic acid with sodium hydroxide?

(*iii*) How are the figures for the 'useful pH range' obtained?

Comment on the following observation. Addition of a small quantity of bromocresol green to 0.1 M aqueous ammonium chloride at 298 K results in a blue coloration. If the solution is warmed to 360 K in the absence of air the colour changes to green; the blue colour is restored on cooling.

(*Oxford*)

17 (*a*) Define an acid and a base *in terms of proton transfer*. Give one example in each case.

(*b*) The self-ionization of liquid sulphur dioxide can be represented by the equation:

$$2SO_2(l) \rightleftharpoons SO^{2+}(\text{in liquid } SO_2) + SO_3^{2-} \text{ (in liquid } SO_2).$$

Use this equation to explain what is meant by (*i*) a Lewis acid; (*ii*) a Lewis base.

(*c*) Anhydrous aluminium chloride is a Lewis acid which is used in the alkylation of benzene. It gives rise to an electrophile, CH_3^+, with chloromethane, CH_3Cl. Write an equation to show the formation of the CH_3^+, stating how the anhydrous aluminium chloride acts as a Lewis acid.

(*AEB 1980*)

18 (*a*) Define the terms "Lewis acid" and "Lewis base" and give one example of each.

(*b*) Define the terms "acid" and "base" on the Brønsted-Lowry theory and give one example of each.

(*c*) Explain why

(*i*) CCl_3COOH is a stronger acid than CH_3COOH.

(*ii*) the $[Al(H_2O)_6]^{3+}$ ion is a stronger acid than the $[Mg(H_2O)_6]^{2+}$ ion.

(*JMB*)

19 In a buffered solution

$$pH = -\lg K_a + \lg \frac{[\text{base}]_{eqm}}{[\text{acid}]_{eqm}}$$

Human plasma is buffered mainly by dissolved carbon dioxide which has reacted to form carbonic acid.

$$H_2CO_3(aq) \rightleftharpoons H^+(aq) + HCO_3^-(aq)$$

(*a*) Explain how carbonic acid can buffer human plasma. Give an example to illustrate your answer.

(*b*) When the concentrations of carbonic acid and hydrogencarbonate ion are equal, the concentration of hydrogen ions is $7.9 \times 10^{-7} \text{ mol dm}^{-3}$. Calculate the value of $\lg K_a$ for carbonic acid.

(*c*) Usually the pH of human plasma is about 7.4. Calculate the *ratio* of the concentrations of hydrogencarbonate ion and carbonic acid in plasma.

(*d*) If the total concentration of hydrogencarbonate ion and carbonic acid was equivalent to $2.52 \times 10^{-2} \text{ mol dm}^{-3}$ of carbon dioxide, calculate the separate *concentrations* of the hydrogencarbonate ion and the carbonic acid in plasma.

(*Nuffield*)

Principles of physical chemistry

20 Explain what is meant by pH and describe briefly one method of measuring it.

A weak acid HA has an ionization constant $K_a = 2 \times 10^{-5}$ mol dm^{-3}. A solution S is made up with 0.1 mole of HA in 1 dm^3, and a second solution P has 0.1 mole of HA and 0.5 mole of the sodium salt of the weak acid (NaA) in 1 dm^3. Calculate the pH of solutions S and P, stating any approximations that are made. If 0.01 mole of a strong acid HX is added to solution S, and an identical amount is added to solution P, what are the new pH values? Comment on your results.

(Oxford Entrance)

13

Chemical energy and electrical energy: electrochemistry

The generation and storage of electrical power in mobile sources – in electrical cells, fuel cells, and nerve cells – is a central feature of everyday life. The physical chemistry related to them is the subject of this chapter. We shall see that electrons being transferred between species in the course of chemical reactions can be intercepted and used to do work. Measurements of the potential differences across the terminals of cells can be interpreted in terms of thermodynamic properties, and we shall see that information of this kind can be used to predict equilibrium constants for reactions, measure pH, follow titrations, and account for the ability of one metal to displace another from solution.

Introduction

Modern electrochemistry deals with a wide range of important phenomena involving the movement of electrons from one place to another. It grew out of investigations of electrical power production in cells. That remains a major field of study, but it is now much more widely applied. Now electrochemistry includes the fuel cells which are already used in space vehicles and which are being investigated with the aim of developing pollution-free transport systems. It also includes devices capable of turning solar radiation directly into electrical power. Electrochemistry also goes on inside us as well as in our environment. Nerve signals, for example, arise from electrochemical processes pushing spikes of potential difference through the elaborate circuitry of the nervous system and the brain. Hence even to think about electrochemistry involves doing electrochemistry!

Electrochemical measurements are also very useful for investigating the properties of materials and can be used to obtain measurements of equilibrium constants, enthalpies, entropies, and Gibbs functions of reactions. They can also be used analytically, and provide readily automated techniques for analyzing mixtures and following titrations.

13.1 Electrochemical cells

When a piece of zinc is immersed in a solution of copper(II) sulphate, the

Principles of physical chemistry

overall reaction is

$$Zn(s) + CuSO_4(aq) \rightleftharpoons ZnSO_4(aq) + Cu(s);$$
$$K_c = \{[ZnSO_4]/[CuSO_4]\}_{eq} = \{[Zn^{2+}]/[Cu^{2+}]\}_{eq}. \qquad (13.1.1)$$

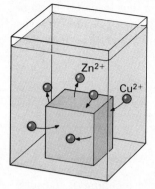

13.1 Zinc metal displacing copper ions

Copper is deposited on the surface of the zinc, which crumbles and goes into solution. The reaction can be thought of as proceeding in two stages, or *half-reactions*. In one, the zinc donates two electrons; $Zn \rightarrow Zn^{2+} + 2e^-$. In the other copper ions in the solution accept the electrons; $Cu^{2+} + 2e^- \rightarrow Cu$. The equilibrium constant for the overall reaction lies strongly in favour of the right-hand side, and the metallic zinc dissolves virtually completely.

The first question this raises is why the equilibrium lies in that direction. The answer lies in zinc's ability to discard electrons; it loses them more easily than does copper, and so when we think of the balance of the two half-reactions, the zinc donation dominates. Zinc can acquire a positive electric charge more readily than can copper because its valence electrons are more easily removed, and so we say that it is more *electropositive* than is copper. This suggests that it may be possible to set up a measure of electropositive character and use it to predict the positions of equilibria.

As depicted in Fig. 13.1, the reaction takes place as electrons are transferred in random directions between the metal and the solution. But suppose we could find a way of letting the electrons dump into an electrode, travel through an external circuit, and then return to the system through another electrode, where they can attach to the copper ions, Fig. 13.2. The reaction is exactly the same as before, but as it takes place electrons flow through an external circuit and can be used to do useful work, such as driving an electric motor, lighting an electric bulb, or driving some other electrical equipment. This is in fact the basis of the operation of electric batteries and fuel cells, and the zinc/copper reaction is used in the *Daniell cell*. The familiar *dry cell*, Fig. 13.3, is powered by the reaction in which the zinc case supplies electrons through the external circuit and back *via* the central electrode to the manganese(IV) oxide in the interior of the cell.

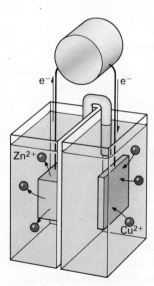

13.2 Zinc metal displacing copper ions with a flow of electrons through an external circuit

The ideas in the last two paragraphs are linked. Not only shall we be able to set up a scale of electropositive character, but we shall also be able to use it to discuss the generation of electricity by chemical reactions.

Redox reactions

We have seen the usefulness of generalizing the simple acid–base concept. We shall now introduce a definition that generalizes another familiar concept. This brings under the same umbrella a wide variety of reactions that are important in electrochemistry.

A simple version of an *oxidation* reaction is combustion, in which some material reacts with oxygen. For instance, zinc reacts with oxygen according to

$$Zn(s) + \tfrac{1}{2}O_2(g) \rightarrow ZnO(s).$$

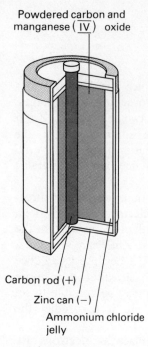

Powdered carbon and
manganese (IV) oxide

Carbon rod (+)

Zinc can (−)

Ammonium chloride
jelly

13.3   A dry cell

(a)

e^-

Oxidation

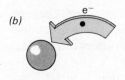

(b)

e^-

Reduction

**13.4   (a) Oxidation and
(b) reduction**

A *formal* way of expressing the reaction (*i.e.*, one that leads to the same reaction overall, but not signifying how it actually takes place) is in terms of two steps, one the release of electrons by zinc, and the other their acceptance by oxygen:

$$Zn \rightarrow Zn^{2+} + 2e^-, \quad \tfrac{1}{2}O_2 + 2e^- \rightarrow O^{2-}.$$

This suggests a broadening of the concept of oxidation. From now on we mean the following, Fig. 13.4(*a*):

Oxidation: the removal of electrons from a species.

This generalization obviously captures the simple meaning of the term oxidation, but it also includes reactions in which oxygen is not involved. For instance, the conversion of iron(II) to iron(III) is formally $Fe^{2+}(aq) \rightarrow Fe^{3+}(aq) + e^-$ and so is an oxidation in the general sense.

The other type of reaction we shall generalize is *reduction*. In the early days of chemistry a reduction was a reaction with hydrogen. For instance, when hydrogen is passed over hot copper(II) oxide the following reaction takes place:

$$CuO(s) + H_2(g) \rightarrow Cu(s) + H_2O(g).$$

This reaction can *formally* be broken into two half-reactions:

$$H_2 \rightarrow 2H^+ + 2e^-, \quad Cu^{2+} + 2e^- \rightarrow Cu$$

and the $2H^+$ can be thought of as combining with the O^{2-} of CuO to produce water. This suggests a useful way of broadening the concept of reduction. From now on we mean the following, Fig. 13.4(*b*):

Reduction: The donation of electrons to a species.

This generalization captures the simple meaning of the term reduction, but it also includes reactions in which hydrogen is not involved. As an example we have the reverse of the iron(II) oxidation: $Fe^{3+}(aq) + e^- \rightarrow Fe^{2+}(aq)$. The reverse of any oxidation is a reduction.[1]

If we return to the zinc/copper reaction that introduced this section, we can formally regard it as the coupling of a reduction reaction ($Cu^{2+} + 2e^- \rightarrow Cu$) with an oxidation reaction ($Zn \rightarrow Zn^{2+} + 2e^-$). It is therefore called a *redox reaction*. The Cu^{2+} is the *oxidizing agent* (which is itself reduced) while the Zn is the *reducing agent* (and is itself oxidized). All the reactions that we are going to consider can be regarded as redox reactions. Reactions used to generate electricity involve redox systems, with the reducing and oxidizing half-reactions taking place at opposite ends of the external circuit, Fig. 13.2. Redox reactions also feature in other fields – 'breathalyzers', for example, depend on the redox reaction between alcohol and potassium dichromate(VI), which turns the yellow salt green.

---

[1] One way of remembering which is which is OIL RIG: Oxidation Is Loss, Reduction Is Gain (of electrons).

Principles of physical chemistry

Electrode potentials

A typical arrangement for capturing the electrical driving power of chemical reactions is the *electrochemical cell* depicted in Fig. 13.5. It consists of two *half-cells* in electrical contact through a conducting medium known as a *salt bridge*. When an external circuit is completed, as in Fig. 13.6, the zinc electrode donates electrons, and in the process generates Zn^{2+} ions, which break away from the metal (by hydration) and drift off into the solution. The electrons pass through the external circuit and enter the copper electrode. There they attach to Cu^{2+} ions in the neighbouring solution, reducing them to Cu atoms, which stick to the electrode.

Electrons, being negatively charged, tend to travel towards regions of positive electric potential. The observation that the electrons travel from the zinc electrode to the copper therefore tells us that the copper electrode is at a *positive* electrical potential relative to the zinc electrode, Fig. 13.6.

The next step is to measure the magnitude of the potential difference between the two electrodes. This has to be done without drawing current so that the concentrations of the solutions in the half-cells do not change (for otherwise the potential difference would change, and there would be other complicating effects). In modern measurements a high-resistance digital voltmeter is used.

The potential difference measured between the electrodes when negligible current is being drawn is called the *electromotive force* (e.m.f.) of the cell. The e.m.f. of the present cell might be about 1.14 V, depending on

Salt bridge

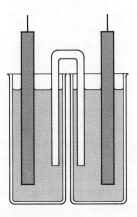

or, more simply :

13.5 An electrochemical cell

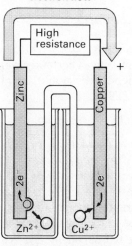

Tendency of electron flow

High resistance

13.6 The relative electric potentials of electrodes and the tendency of electron flow

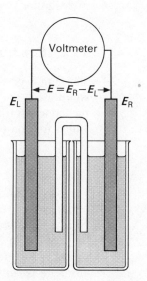

Voltmeter

$E = E_R - E_L$

E_L E_R

13.7 The sign convention for a cell e.m.f.

Low electrode potential   High electrode potential

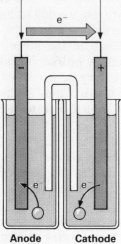

Anode
(site of oxidation)

Cathode
(site of reduction)

13.8 The definition of anode and cathode, and the signs of the electrode potentials

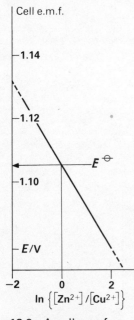

13.9 A cell e.m.f. depends logarithmically on the ion concentrations

the concentration and the temperature. We could also report that the copper electrode is positive relative to the zinc.

In order to avoid having to state which is the positive electrode every time an e.m.f. is reported, the following convention is used. A cell is denoted by a series of compartments. In the present case we have

$$Zn(s) \mid ZnSO_4(aq) \mid CuSO_4(aq) \mid Cu(s).$$

The junctions between different media are denoted by single vertical lines. The concentrations can also be specified. The measured e.m.f. is then written beside it, using the convention that its value is *the potential of the right-hand electrode minus the potential of the left-hand electrode*, Fig. 13.7.

E.m.f.: $E(cell) = Potential(r.h.\ electrode) - Potential(l.h.\ electrode),$

it being understood that no current is drawn in the course of the measurement. Therefore, if the right-hand *electrode potential* is higher than that of the left, the e.m.f. has a positive sign; if it is lower, the e.m.f. has a negative sign. In the present case we can write

$$Zn(s) \mid ZnSO_4(aq, 0.01\ mol\ dm^{-3}) \mid CuSO_4(aq, 0.2\ mol\ dm^{-3}) \mid Cu(s);$$
$$E = +1.14\ V.$$

If we had written the cell in the reverse order, we would have specified the e.m.f. as -1.14 V.

The important point to remember is that *the electrode with the higher potential is where reduction occurs* and *the electrode with the lower potential is where oxidation occurs*, Fig. 13.8. The former is called the *cathode* (the site of reduction, by definition), and the latter is called the *anode* (the site of oxidation).

13.2   E.m.f. and equilibria

The e.m.f. of the zinc/copper cell depends on the concentrations of the solutions in the two half-cells. Thermodynamic considerations predict that the concentration dependence should be given by the *Nernst equation*, which for the present cell is

$$E = E^{\ominus} - (RT/2F) \ln \{[Zn^{2+}]/[Cu^{2+}]\}, \tag{13.2.1}$$

where R is the molar gas constant, T the absolute temperature, and F the Faraday constant. $E^{\ominus}$, the *standard e.m.f.* of the cell, is the value of the e.m.f. when the concentrations of the solutions are 1 mol dm^{-3}. The Nernst equation suggests that if the e.m.f. of the cell is plotted against $\ln \{[Zn^{2+}]/[Cu^{2+}]\}$ a straight line should be expected. This is observed in practice, Fig. 13.9. The standard e.m.f. is the difference of two *standard electrode potentials* (or *redox potentials*). We can write

$$E^{\ominus}(cell) = E^{\ominus}(r.h.\ electrode) - E^{\ominus}(l.h.\ electrode). \tag{13.2.2}$$

The standard electrode potential is the potential of the electrode when the concentration of the neighbouring solution is 1 mol dm^{-3}.

Principles of physical chemistry

Now consider the following question. What does it mean to say that a cell is *exhausted*? An exhausted cell is one that has no further tendency to generate an electric current through an external circuit (*e.g.*, *a* 'flat' battery). That condition corresponds to the cell reaction having no further tendency to occur. In other words, the cell reaction is at equilibrium. Therefore, *when the reactants have their equilibrium concentrations, the e.m.f. is zero*. In the case of the zinc/copper cell we can therefore write

$$0 = E^{\ominus} - (RT/2F) \ln \{[Zn^{2+}]/[Cu^{2+}]\}_{eq}.$$

The ratio of concentrations in this expression is nothing other than the equilibrium constant for the cell reaction, eqn (13.1.1). Therefore, for this reaction,

$$0 = E^{\ominus} - (RT/2F) \ln K_c, \quad \text{or} \quad (RT/2F) \ln K_c = E^{\ominus}. \tag{13.2.3}$$

The importance of this result cannot be overemphasized; simply by looking at a table of standard electrode potentials, and combining their values to form $E^{\ominus}$ for the reaction of interest, we can predict the value of the equilibrium constant for the reaction.

Table 13.1 Standard electrode potentials at 25 °C, $E^{\ominus}/V$

Most electropositive, most reducing	
$Li^+ + e^- \rightleftharpoons Li$	−3.045
$K^+ + e^- \rightleftharpoons K$	−2.924
$Ca^{2+} + 2e^- \rightleftharpoons Ca$	−2.76
$Na^+ + e^- \rightleftharpoons Na$	−2.712
$Mg^{2+} + 2e^- \rightleftharpoons Mg$	−2.375
$Al^{3+} + 3e^- \rightleftharpoons Al$	−1.706
$Zn^{2+} + 2e^- \rightleftharpoons Zn$	−0.763
$Fe^{2+} + 2e^- \rightleftharpoons Fe$	−0.409
$Sn^{2+} + 2e^- \rightleftharpoons Sn$	−0.136
$Pb^{2+} + 2e^- \rightleftharpoons Pb$	−0.126
$2H^+ + 2e^- \rightleftharpoons H_2$	0
$AgBr + e^- \rightleftharpoons Ag + Br^-$	0.071
$AgCl + e^- \rightleftharpoons Ag + Cl^-$	0.223
$Cu^{2+} + 2e^- \rightleftharpoons Cu$	0.340
$Cu^+ + e^- \rightleftharpoons Cu$	0.522
$I_2 + 2e^- \rightleftharpoons 2I^-$	0.535
$Fe^{3+} + e^- \rightleftharpoons Fe^{2+}$	0.770
$Hg_2^{2+} + 2e^- \rightleftharpoons 2Hg$	0.799
$Ag^+ + e^- \rightleftharpoons Ag$	0.800
$2Hg^{2+} + 2e^- \rightleftharpoons Hg_2^{2+}$	0.905
$Br_2 + 2e^- \rightleftharpoons 2Br^-$	1.065
$4H^+ + O_2 + 4e^- \rightleftharpoons 2H_2O$	1.229
$Cl_2 + 2e^- \rightleftharpoons 2Cl^-$	1.358
$Au^{3+} + 3e^- \rightleftharpoons Au$	1.42
$Ce^{4+} + e^- \rightleftharpoons Ce^{3+}$	1.443
Least electropositive, least reducing	

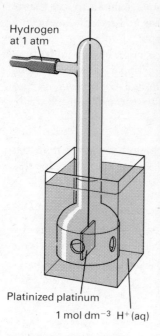

Hydrogen at 1 atm

Platinized platinum

1 mol dm^{-3} H$^+$(aq)

13.10 The standard hydrogen electrode

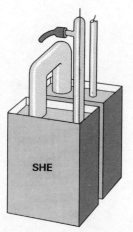

SHE

13.11 Measuring a standard electrode potential (of the cell on the right)

Example: Finely divided lead and tin were shaken with solutions containing tin(II) and lead(II) perchlorates until the equilibrium $Sn(s) + Pb^{2+}(aq) \rightleftharpoons Sn^{2+}(aq) + Pb(s)$ was reached. What is the equilibrium constant of the reaction?

Method: Use eqn (13.2.3). Find $E^{\ominus}$ from $E^{\ominus} = E^{\ominus}(Pb^{2+}, Pb) - E^{\ominus}(Sn^{2+}, Sn)$ using data from Table 13.1. At 25°C, $RT/F = 0.02569$ V.

Answer: $E^{\ominus} = (-0.126 \text{ V}) - (-0.136 \text{ V}) = 0.010 \text{ V}$. Then from eqn (13.2.3),

$$\ln K_c = \frac{2 \times (0.010 \text{ V})}{(0.02569 \text{ V})} = 0.779.$$

The equilibrium constant for the reaction is therefore

$$K_c = \left\{ \frac{[Sn^{2+}]}{[Pb^{2+}]} \right\}_{eq} = 2.2.$$

Comment: Notice that tin, with the slightly more negative electrode potential, is present in higher concentration at equilibrium. Eqn (13.2.3) can be used when, as in this reaction, every redox step involves two electrons. We treat the general case later.

Standard electrode potentials

In order to determine standard electrode potentials we have to measure the e.m.f. of a cell in which the concentrations of the solutions are all 1 mol dm^{-3}. An immediate problem, however, is that every measurement gives only the difference of electrode potentials. This problem is solved by selecting one electrode potential and *defining* it to be zero. Since every application of electrode potentials depends on differences, this introduces no difficulties of any practical importance.

The electrode chosen to have $E^{\ominus} = 0$ is the *standard hydrogen electrode* (SHE). It is prepared as shown in Fig. 13.10. The equilibrium established when it is in operation is

$$H^+(aq) + e^- \rightleftharpoons \tfrac{1}{2} H_2(g).$$

The standard value $E^{\ominus}(SHE) = 0$ refers to its potential when the hydrogen ion concentration is 1 mol dm^{-3} (*i.e.*, pH = 0) and the hydrogen gas is in its standard state (1 atm pressure at the specified temperature, which is usually taken as 25 °C). Any other standard electrode potential can be obtained by measuring the standard e.m.f. of a cell constructed with the SHE on the left and the electrode of interest on the right, Fig. 13.11:

$$E^{\ominus}(cell) = E^{\ominus}(r.h. \text{ electrode}) - E^{\ominus}(SHE) = E^{\ominus}(r.h. \text{ electrode})$$

because $E^{\ominus}(SHE) = 0$. In practice a SHE is inconvenient to set up and use, and so *secondary standards* are used. These are electrodes with well-established properties calibrated against a SHE. The most widely used is the *calomel electrode*, which is constructed as indicated in Fig. 13.12 ('calomel' is the trivial name for mercury(I) chloride, Hg_2Cl_2); its standard potential is 0.26 V at 25 °C. Values of standard electrode potentials are listed in Table 13.1.

13.3 Using electrode potentials

Cell e.m.f. measurements, and the tables of standard electrode potentials they lead to, have a wide range of applications. We shall consider three important examples, but before going on it is interesting to note that dental fillings can be used as the source of electric current. A filling is typically an amalgam of silver and tin in mercury. When aluminium foil is bitten a cell is set up in which the electrolyte is saliva, the aluminium is the anode, and the filling is the cathode. The e.m.f. is about 2 V, enough to cause discomfort but not death. A mixture of gold and amalgam fillings can lead to a persistent metallic taste because tin ions are released from the amalgam, which now is the anode.

Calculating equilibrium constants

The basic equation for applying standard electrode potentials is their connection with the equilibrium constant. Eqn (13.2.3) is used when the redox reaction involves two electrons, but in cases where z electrons are involved the 2 in $RT/2F$ has to be replaced by z. Therefore the general expression is

$$\ln K_c = zFE^{\ominus}/RT = zE^{\ominus}/(0.0257 \text{ V}) \quad \text{at } 25 \text{°C}. \tag{13.3.1}$$

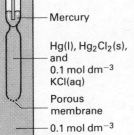

- Platinum wire

- Mercury

Hg(l), Hg$_2$Cl$_2$(s), and 0.1 mol dm^{-3} KCl(aq)

Porous membrane

0.1 mol dm^{-3} KCl(aq)

Porous membrane

13.12 A calomel electrode

Example: Predict the value of the equilibrium constant for the reaction $Zn(s) + CuSO_4(aq) \rightleftharpoons ZnSO_4(aq) + Cu(s)$ at 25 °C.

Method: The equilibrium constant is

$$K_c = \left\{\frac{[Zn^{2+}][SO_4^{2-}]}{[Cu^{2+}][SO_4^{2-}]}\right\}_{eq} = \left\{\frac{[Zn^{2+}]}{[Cu^{2+}]}\right\}_{eq}$$

because the sulphate concentrations cancel. Use eqn (13.3.1) with $z=2$ in order to find K_c. Find $E^{\ominus}$ from $E^{\ominus} = E^{\ominus}(Cu^{2+}, Cu) - E^{\ominus}(Zn^{2+}, Zn)$; values are listed in Table 13.1.

Answer: Since $E^{\ominus}(Cu^{2+}, Cu) = 0.340$ V and $E^{\ominus}(Zn^{2+}, Zn) = -0.763$ V the overall cell e.m.f. is $E^{\ominus} = (0.340 \text{ V}) - (-0.763 \text{ V}) = 1.103$ V. Eqn (13.3.1) then gives

$$\ln K_c = \frac{2 \times (1.103 \text{ V})}{(0.0257 \text{ V})} = 85.8; \quad \text{therefore} \quad K_c = 1.9 \times 10^{37}.$$

Comment: Note that we have avoided the need to perform an experiment by making use of tabulated data (although, of course, Table 13.1 has been compiled by a lot of people doing a lot of careful experiments). Table 13.1 can be used to discuss the equilibrium positions of a wide range of reactions, but note that it applies to 25 °C.

When the value of $E^{\ominus}$ for the cell reaction is large and positive the equilibrium constant is also large, and so the equilibrium lies strongly in favour of the products. In contrast, if $E^{\ominus}$ is negative, then the equilibrium lies in favour of the reactants. An easy way of remembering this is

$E^{\ominus} > 0$ favours $\rightarrow$; $\qquad E^{\ominus} < 0$ favours $\leftarrow$.

Displacing, reducing, and corroding

We can now set up a scale of electropositive character, an *electrochemical series*. This lets us say at a glance whether one metal will displace another from solution (as zinc displaces copper).

A typical displacement reaction is

$$Zn(s) + Cu^{2+}(aq) \rightleftharpoons Zn^{2+}(aq) + Cu(s).$$

The corresponding standard e.m.f. is $E^{\ominus} = 1.103$ V (see the *Example*). This is *positive*, the equilibrium constant is large (1.9×10^{37}), the products are favoured, and zinc displaces copper from solution. Since Zn tends to form Zn^{2+} more than Cu tends to form Cu^{2+}, we conclude that zinc is more *electropositive* than is copper. In general, the *more negative* the standard electrode potential, the *more electropositive* the metal. Since a more electropositive metal has a tendency[1] to displace a less electropositive metal, *a metal of a given standard electrode potential has a tendency to displace any metal with a greater (more positive) standard electrode potential.*

Example: Can aluminium displace copper from solution?

Method: Inspect Table 13.1 to see whether aluminium is more electropositive than copper; if it is it can. Table 13.1 is arranged so that electropositive character increases upwards, therefore inspect it to see whether aluminium lies above copper.

Answer: Aluminium lies above copper, and so aluminium has a thermodynamic tendency to displace copper from solution.

Comment: The difference of the electrode potentials, 2.046 V, is enormous, and corresponds to the equilibrium constant for the displacement reaction having a value of around 10^{200}. The displacement therefore has a tendency to go to virtual completion. Nevertheless, this equilibrium might be reached only very slowly, because aluminium left open to the air acquires an impenetrable layer of oxide. This underlines the difference between a *tendency* to react and the *rate* of the reaction.

[1] The *tendency* of the displacement to occur has to be understood as meaning that the *equilibrium constant* for the displacement favours products. Whether in practice this means that one metal goes into solution at the expense of the other, or vice versa, depends on the starting concentrations. For instance, although lead is less electropositive than tin, when a lump of lead is dropped into a tin solution it displaces the tin until the equilibrium constant $K_c = 2.2$ is reached. When $E^{\ominus}$ is large (greater than about 0.1 V) this is an academic point. For example, if a lump of copper is dropped into a 1 km³ lake of 1 mol dm⁻³ solution of a zinc salt only one atom of copper needs to go into solution before the equilibrium constant (10^{37}) is reached.

The behaviour we have described can be generalized as follows. The zinc/copper displacement is a reduction; the displacing metal (Zn) reduces the metal displaced (Cu^{2+}). Therefore we can say that *a highly electropositive element reduces a less electropositive element*. That is, a species with *low* standard electrode potential has a tendency to reduce a species with a *high* electrode potential. The electrochemical series in Table 13.1 is therefore also arranged with *reducing power* increasing upwards.

An important example of reduction is *corrosion*. In the case of steel objects the reducing agent is iron, which in the process is itself oxidized; the iron salts so formed, especially the oxide, lack the form and rigidity of the original metal structure. One of the reactions the iron drives is the reduction of oxygen. In an acid environment (as in an industrial area) the reaction is

$$4H^+ + O_2 + 4e^- \rightleftharpoons 2H_2O.$$

Since this reaction lies lower down the electrochemical series than iron, Table 13.1, the latter has the greater reducing power and tends to drive it. Therefore, in a moist acidic environment iron tends to form Fe^{2+}, and corrodes. There are ways of protecting iron objects, such as galvanizing and tin plating, and they can also be discussed in terms of the electrochemical series.

Example: Explain the electrochemistry involved in (*a*) galvanizing and (*b*) tin-plating iron objects.

Method: 'Galvanizing' means zinc-plating. Note the relative position of the metals in the electrochemical series, Table 13.1. If the plating metal is more electropositive than iron it tends to reduce any iron ions that may be present, and therefore tends to conserve it as the metal.

Answer: (*a*) Zinc is more electropositive than iron, and therefore tends to reduce any iron ions that form. Its presence therefore prevents corrosion. (*b*) Iron is more electropositive than tin, and so it will supply electrons to any tin ions that are present. Therefore, once tin ions are available, tin *aids* the oxidation of iron.

Comment: When the surface of a galvanized iron object is broken the zinc tends to corrode and the iron not. The object keeps its form at the expense of the zinc. Tin-plating protects the iron from attack by moist air, but once the surface is broken the iron can reduce any tin ions that are present. A scratched tin-plated object therefore has a strong tendency to corrode, and once damaged it rusts extensively.

Measuring pH

The measurement of e.m.f. gives a direct and simple method of determining pH. We have already seen that electrode potentials depend on the logarithms of concentrations. The hydrogen electrode, in particular, has a

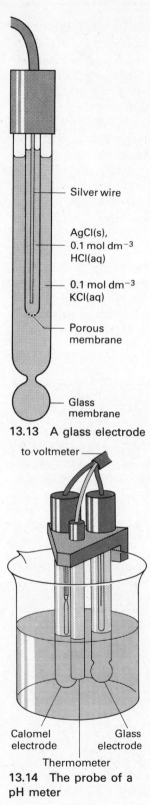

Silver wire

AgCl(s),
0.1 mol dm^{-3}
HCl(aq)

0.1 mol dm^{-3}
KCl(aq)

Porous
membrane

Glass
membrane

13.13 A glass electrode

to voltmeter

Calomel
electrode

Glass
electrode

Thermometer

**13.14 The probe of a
pH meter**

potential that depends on the logarithm of the hydrogen ion concentration, in other words, on the pH. Therefore, all that it is necessary to do (in principle) is to set up a cell with a hydrogen electrode dipping into the solution of interest.

In practice there is a much easier technique. The hydrogen electrode is messy and tiresome to set up, but it can be replaced by any other electrode that responds to the hydrogen ion concentration. One such is the *glass electrode*, Fig. 13.13, which has a potential proportional to the pH. It is normally used in conjunction with a calomel electrode, and the e.m.f. of the complete cell, Fig. 13.14, is read with a high-resistance voltmeter. Since the e.m.f. is proportional to the pH, the scale can be marked so as to display the pH directly. Since in this way pH is represented by an electrical signal, titrations can be performed entirely automatically and the analysis dealt with on a microprocessor linked to the equipment. Apart from being accurate and reliable, automation of this kind releases the research worker (and the student) for more imaginative and interesting investigations.

13.4 The thermodynamics of e.m.f.

In this section we shall demonstrate two things. One is the way in which measurement of standard e.m.f. can be used to determine the standard Gibbs function of reaction, $\Delta G^{\ominus}$. We shall also indicate the thermodynamic reason why e.m.f. and Gibbs function are related, and use the result to discuss power generation.

Measuring $\Delta G^{\ominus}$

We have met two equations, one being the relation between $\Delta G^{\ominus}$ and the equilibrium constant, eqn (11.3.3), and the other the relation between the equilibrium constant and standard e.m.f., eqn (13.3.1). For convenience we write them again here:

$$\Delta G_m^{\ominus} = -RT \ln K_c \quad \text{and} \quad RT \ln K_c = zFE^{\ominus}.$$

They obviously combine together by elimination of $RT \ln K_c$. This gives

$$\Delta G_m^{\ominus} = -zFE^{\ominus}. \tag{13.4.1}$$

Simply by measuring $E^{\ominus}$ of a cell we can find the value of $\Delta G_m^{\ominus}$ for a reaction. Alternatively, knowing $\Delta G_m^{\ominus}$ we can predict the standard e.m.f. of the corresponding cell.

Example: What is the standard molar Gibbs function of the reaction $Zn(s) + CuSO_4(aq) \rightleftharpoons ZnSO_4(aq) + Cu(s)$ at 25 °C?

Method: Use eqn (13.4.1). Since two electrons are transferred in the reaction take $z = 2$. Take the standard electrode potentials from Table 13.1. Use $1 \, C \, V = 1 \, J$.

Principles of physical chemistry

Answer: The standard e.m.f. corresponding to the cell reaction is

$$E^{\ominus} = E^{\ominus}(Cu^{2+}, Cu) - E^{\ominus}(Zn^{2+}, Zn)$$
$$= (0.340\ V) - (-0.763\ V) = 1.103\ V.$$

Since $z = 2$, eqn (13.4.1) gives

$$\Delta G_m^{\ominus} = -2 \times (9.648 \times 10^4\ C\ mol^{-1}) \times (1.103\ V)$$
$$= -2.128 \times 10^5\ C\ V\ mol^{-1} = -212.8\ kJ\ mol^{-1}.$$

Comment: Note that this is the *standard* molar Gibbs function at 25 °C: it corresponds to the change in the Gibbs function when zinc displaces copper ions from a $1\ mol\ dm^{-3}$ solution of $CuSO_4(aq)$ to form a $1\ mol\ dm^{-3}$ $ZnSO_4(aq)$ solution, there being no Zn^{2+} ions in the initial solution, and no Cu^{2+} ions in the final solution (*i.e.*, 'complete reaction'). Note that $\Delta G_m^{\ominus}$ is strongly negative, and so the reaction has a strong tendency to run to the right. This is a thermodynamic way of explaining why zinc displaces copper.

Work, cells, and fuel cells

In Section 11.3 we saw that the change in Gibbs function in the course of a reaction is equal to the maximum electrical work the reaction can produce: $\Delta G = w_{e,max}$. The importance of this relation in practical affairs is as follows. The construction of a typical *fuel cell* is shown in Fig. 13.15; it operates exactly like an electrochemical cell, electrons being dumped in one electrode and travelling through an external circuit to bring about reduction at the other, the only difference being that the fuel for the reaction is provided from external reserves of gas. If we know the value of ΔG for the reaction, we can state without further calculation the maximum electrical work the cell can supply. In the case of the hydrogen/oxygen fuel cell the reaction is

$$2H_2(g) + O_2(g) \rightarrow 2H_2O(l);$$
$$\Delta G_m^{\ominus} = 2 \times (-237.2\ kJ\ mol^{-1}) = -474.4\ kJ\ mol^{-1} \quad \text{at } 25\ °C.$$

($\Delta G_m^{\ominus}$ comes from Table 11.2.) We therefore know at once that for every 36 g of water produced, the cell can generate 474.4 kJ of electrical work when the gases are at pressures of 1 atm and the temperature is 25 °C. This is enough to raise a 1 tonne object through about 50 m.

This calculation can be taken further because we can also predict the standard e.m.f. of the fuel cell. All we have to do is to use eqn (13.4.1) to convert the value of $\Delta G_m^{\ominus}$ to a value of $E^{\ominus}$. Since $z = 4$ for the reaction we find $E^{\ominus} = 1.23$ V. Calculations like this are at the heart of modern work on the development of fuel cells.

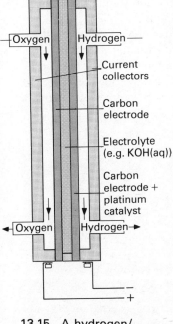

Oxygen — Hydrogen

Current collectors

Carbon electrode

Electrolyte (e.g. KOH(aq))

Carbon electrode + platinum catalyst

Oxygen — Hydrogen

13.15 A hydrogen/oxygen fuel cell element

Summary

1 *Oxidation* is the removal of electrons from a species; *reduction* is the donation of electrons to a species.

2   *Redox reactions* are combinations of oxidation and reduction reactions; the *oxidizing agent* is itself reduced and the *reducing agent* is itself oxidized.

3   An *electrochemical cell* consists of two *half-cells*; in some cases they are joined through a *salt bridge* which establishes electrical contact.

4   The *electromotive force* (*e.m.f.*) of a cell is the electric potential difference between the electrodes of the cell, $E(\text{right}) - E(\text{left})$, when no current is drawn. The e.m.f. is measured with a high-resistance digital voltmeter.

5   The e.m.f. of a cell depends on the logarithm of the *concentrations* of the species present; the e.m.f. of the cell when the concentrations of the solutions are $1\,\text{mol dm}^{-3}$ is the *standard e.m.f.*, $E^{\ominus}$.

6   The standard e.m.f. of a cell is the difference between the *standard electrode potentials* (or *redox potentials*) of the two half-cells.

7   Standard electrode potentials are reported on the basis that the potential of the *standard hydrogen electrode* is zero. In practice the *calomel electrode* is used as a secondary standard.

8   When a cell reaction is at *equilibrium* it produces no e.m.f.

9   The equilibrium constant K_c and the standard e.m.f. of a reaction are related by $RT \ln K_c = zFE^{\ominus}$.

10   Standard electrode potentials determine the *electrochemical series*; a metal of a given standard electrode potential will be displaced or reduced by a metal higher up the series (with a lower standard electrode potential).

11   A *pH meter* makes use of the *glass electrode*, an electrode with a potential that depends on the hydrogen ion concentration of the medium, in combination with a reference electrode.

12   The standard molar Gibbs function $\Delta G_m^{\ominus}$ and the standard electrode potential of a reaction are related by $\Delta G_m^{\ominus} = -zFE^{\ominus}$.

13   Measurements of the e.m.f. of a cell are used to obtain *thermodynamic information*, such as the value of ΔG for the reaction.

14   The maximum *electrical work* that can be obtained in the course of a chemical reaction is $w_{e,max} = \Delta G$; thermodynamic data can therefore be used to assess the usefulness of chemical reactions as sources of electrical energy.

Problems

1   Define the terms *oxidation, reduction,* and *redox reaction*.

2   State which of the following reactions are redox reactions and, in those cases, identify the species being reduced and oxidized.

(a) $Zn + S \rightarrow ZnS$ (b) $H_2S + Cl_2 \rightarrow S + 2HCl$
(c) $KOH + HCl \rightarrow KCl + H_2O$ (d) $CuO + H_2 \rightarrow Cu + H_2O$
(e) $C_2H_4 + HBr \rightarrow C_2H_5Br$ (f) $2H_2O_2 \rightarrow 2H_2O + O_2$
(g) $MnO_2 + 4HCl \rightarrow MnCl_2 + Cl_2 + 2H_2O$ (h) $Fe_2O_3 + 3CO \rightarrow 2Fe + 3CO_2$.

3   Aluminium objects frequently fail to show the reactivity expected on the basis of its position in the electrochemical series. Explain this observation.

4   State which is the positive electrode in the following electrochemical cells:

(a) $Zn \mid ZnSO_4(aq) \mid CuSO_4(aq) \mid Cu$,

(b) $Pt, H_2 \mid HCl(aq) \mid AgCl, Ag$,

(c) $Pt, O_2 \mid HCl(aq) \mid H_2, Pt$,

(d) $Ag, AgCl \mid HCl(aq) \mid HBr(aq) \mid AgBr, Ag$.

5 What are the standard e.m.fs (at 25 °C) of the cells in the last question?

6 Write the cell reactions, expressions for the equilibrium constants of the cell reactions, and deduce the values of the equilibrium constants for the reactions in Question 4.

7 Arrange the following metals in order of decreasing displacing power: aluminium, copper, iron, magnesium, sodium, zinc.

8 Why should copper and iron pipes not be used in the same domestic water system?

9 A lump of magnesium is sometimes buried near a steel object (e.g., a pipeline) and connected to it by conducting cables. Why?

10 Given a molar solution of silver nitrate in water and any equipment necessary, describe how you would determine the standard electrode potential of silver.

Potassium manganate(VII) (KMnO₄) and hydrogen peroxide react in the presence of acid with the evolution of oxygen to give a clear solution. Show how to use the following information to determine the stoichiometric (balanced) ionic equation of the reaction.

$$MnO_4^- + 8H^+ + 5e^- \rightleftharpoons Mn^{2+} + 4H_2O \qquad E^\ominus = +1.52\ V$$

$$O_2 + 2H^+ + 2e^- \rightleftharpoons H_2O_2 \qquad E^\ominus = +0.68\ V$$

What volume of potassium manganate(VII) of concentration $0.200\ mol\ dm^{-3}$ would be required to react with $100\ cm^3$ of hydrogen peroxide of concentration $0.0100\ mol\ dm^{-3}$ (with excess acid present), and what volume of oxygen (at 20 °C and 1 atm pressure) would be evolved?

(Oxford and Cambridge part question)

11 (a) Explain the processes of (i) oxidation, (ii) reduction, in terms of electron transfer.

(b) For each of the following redox reactions, write two balanced equations each involving electron transfer, one for the oxidation process and one for the reduction process (stating in each case which equation describes oxidation and which reduction):

(i) reaction between $Cr_2O_7^{2-}$ and Fe^{2+} in aqueous solution in the presence of H^+;

(ii) reaction between Cl_2 and I^- in aqueous solution;

(iii) reaction between Li and H_2.

(Oxford)

12 (a) Define the standard electrode potential of a metal.

(b) Using the data in Table 13.1,

(i) select two half-cell reactions that may be combined to produce a cell with a standard e.m.f. of +1.10 V. Write an overall equation for the cell reaction and indicate the direction of flow of electrons in a conductor connecting the cell terminals;

(ii) calculate the standard electrode potential for a cell in which the reaction

$$Zn(s) + Fe^{2+}(aq) \rightarrow Zn^{2+}(aq) + Fe(s)$$

occurs and explain the reaction in terms of oxidation and reduction;

(*iii*) explain what happens when copper foil is immersed in aqueous silver nitrate.

(*c*) The dependence of redox potential, *E*, on the ion concentration at 300 K, is given by

$$E = E^{\ominus} + \frac{0.06}{n} \lg \frac{[\text{oxidized form}]}{[\text{reduced form}]}$$

where $E^{\ominus}$ is the standard redox potential and *n* the number of moles of electrons in the half-cell reaction.

(*i*) Using this equation and data from the table above, calculate the equilibrium constant, K_c, for the reaction

$$Ce^{4+}(aq) + Fe^{2+}(aq) \rightleftharpoons Ce^{3+}(aq) + Fe^{3+}(aq)$$

at 300 K. (Note: At equilibrium the potential of each half-cell will be equal.)

(*ii*) From the value that you obtain, comment upon the extent to which iron(II) compounds are oxidized by cerium(IV) compounds in aqueous solution.

(*AEB 1980*)

13   An electrochemical cell

$$Pt(H_2, 1 \text{ atm}) \,|\, HNO_3(m = 1), AgNO_3(m = 1) \,|\, Ag$$

is set up at 298 K. State the e.m.f. of the cell and its polarity on open circuit. What chemical changes begin to occur if it is connected across a high resistance?

The following cell is set up to observe electrolysis at 298 K:

$$Pt(H_2, 1 \text{ atm}) \,|\, HNO_3(m = 1), Cu(NO_3)_2(m = 1) \,|\, Cu.$$

What is the minimum voltage which must be applied, and of what polarity, to cause deposition of copper on the copper electrode? How does the situation differ if:

(*a*) the left-hand electrode is not as shown but is made of Pt with no supply of hydrogen,

(*b*) the left-hand electrode is made of silver with no supply of hydrogen,

(*c*) the electrodes are as shown but the electrolyte is diluted by a factor of ten?

State briefly with reasons whether you expect standard electrode potentials for metal ion/metal electrodes to vary much with temperature.

$Cu^{2+}(aq) \,|\, Cu, E^{\ominus} = 0.337$ V.

$Ag^{+}(aq) \,|\, Ag, E^{\ominus} = 0.799$ V.

(*Cambridge Entrance*)

14

The rate of chemical change

We have seen what holds atoms together in molecules and what determines the tendency of compounds to react. Now we investigate how fast they react. We shall see how to measure rates of reaction, and how to account for their dependence on the composition and the temperature. Some reactions of great importance in industry normally take place far too slowly for commercial use, but we see how they can be made to go faster.

Introduction

Chemical reactions take place at different rates. Some are very slow, like fermentation (which might require several weeks to produce enough product). Others are moderately fast, like the reactions that contract muscles, transmit nervous impulses, and record photographic images. Others are very fast, like the reactions inside the engine of a car. Some are explosively fast.

A knowledge of reaction rates is central to chemistry. In industry it is important to be able to predict how fast a reaction will go under various conditions. Furthermore, many reactions take place as a sequence of simple steps, and we can only truly say that we understand them when the steps have been identified.

14.1 The rates of reactions

Rate in chemistry is defined like speed in physics. It is the change in something divided by the time it takes for the change to occur. The 'something' in this case is the concentration of a reactant (or a product), and so we begin to construct the definition by writing

$$\text{rate of reaction} = \frac{\text{change of concentration of reactant}}{\text{time taken for the change}}.$$

Rate is expressed as 'concentration per unit time'; and since concentrations are often expressed as $mol\ dm^{-3}$ this means that rates are usually quoted as so many $mol\ dm^{-3}\ s^{-1}$.

Measuring the rate
The measurement of rate is like the measurement of speed, and, just as

the speed of a car may change during a journey, so may the rate of a reaction change, see Appendix, p. 231. The concentration of the selected reactant is followed and plotted against time. Then the slope of the graph is evaluated at the time of interest. The slope gives the reaction rate at that time.

The technique used for measuring the concentration depends on how quickly it changes; there is not much point in using a titration method if the reaction is over within a millisecond! For moderately fast reactions, those running to completion in less than about a minute (and including, in modern chemistry, Box 14.1, those requiring as little as 10^{-9} s to run to completion), the best procedure is to use a *spectroscopic method* with electronic recording. For instance, if the reactant of interest is bromine, the intensity of its colour can be interpreted in terms of its concentration. For slower reactions more traditional methods may be used, such as titration. In order to employ these slower techniques a little of the reaction mixture is withdrawn at a series of times, and then diluted very rapidly so that the reaction is *quenched* (slowed to an imperceptible rate). Each sample is then analysed using a suitable titrant or, if the hydrogen ion concentration is required, by measuring its pH. If the reaction produces or removes ions the electrical *conductivity* of the sample changes during the reaction, and so following the sample's conductivity lets us follow the reaction's course. Gas-phase reactions (and reactions producing gases) can often be followed by measuring the *pressure* or the *volume* of gas present.

Box 14.1: The development of chemical kinetics

The first measurement of the rate of a chemical reaction was by L. Wilhelmy (in 1850) who observed the rate at which the optical activity of sucrose solution changed. The concepts of rate law and order of reaction grew out of the work of M. Guldberg and P. Waage (in about 1865) who formulated the *law of mass action*, that 'the rate of a chemical reaction is proportional to the active masses (concentrations) of the reacting substances' (which is what we would now write as rate $= k_2[\text{A}][\text{B}]$). The analysis of many gas-phase reactions was carried out by C. N. Hinshelwood (who received the Nobel Prize, and was simultaneously the President of the Royal Society and of the British Academy), but the delicacy of the problem of working out reaction mechanisms is illustrated by the fact that the mechanism of the decomposition of hydrogen iodide proposed by M. Bodenstein in 1899 survived until it was disproved by J. H. Sullivan in 1967. The development of modern chemical kinetics has been due to the application of electronic methods for following concentrations, and the resolution of changes on ever shorter time scales. Fast reaction techniques include ultrasonic methods, flash photolysis, magnetic resonance, and shock tubes. The

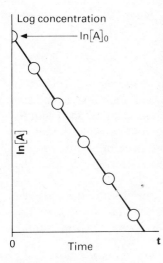

Log concentration

$\ln[A]_0$

$\ln[A]$

0   Time t

14.1 Investigating a first-order reaction

introduction of lasers has extended the time-scale of observations down to about 1 ps $(10^{-12}\,s)$, and extremely detailed information about reactions can now be obtained. Detail of another kind can also now be obtained from *molecular beams*, where streams of molecules are shot at each other and the products are analysed. In this way, the individual events involved in reactions can be understood, and we can now deduce how bonds are made and broken and how molecules are ripped apart or formed during collisions.

At this stage it might appear that all that these experiments are providing are tables of data, and that the only way of talking about reaction rates is to quote the value of the rate for the conditions and times of interest. Happily that is not so. In many cases the reaction rate depends on the concentration in a simple way. For instance, in some reactions it is found that the rate of disappearance of the reactant A is proportional to its concentration, and that rate $\propto [A]$. The discovery of relations like this is a great simplification, and makes the study of reaction rates simple and useful.

14.2 Rate equations

Consider a reaction in which A is destroyed at a rate found to be proportional to its concentration. Then at any stage of the reaction the rate can be written

$$\text{rate} = k_1[A]. \tag{14.2.1}$$

The factor k_1 is the *rate coefficient*, or the *rate constant*; k_1 depends on the conditions, in particular on the temperature, and so the first name is better. This equation is called a *rate equation* or a *rate law*. Because the rate is proportional to the first power of $[A]$ it is termed a *first-order rate equation*. Reactions that show this behaviour are called *first-order reactions* and are said to follow *first-order kinetics*.

First-order reactions

The first advantage of finding (by experiment) that a reaction follows first-order kinetics is that its rate at *any* concentration of A can be summarized by stating the value of the rate coefficient. Thus, if the rate coefficient at some temperature is reported as $2.4 \times 10^{-3}\,s^{-1}$ we would know that the rate of consumption of A at that temperature would be $2.4 \times 10^{-3}\,mol\,dm^{-3}\,s^{-1}$ when the concentration of A is $1.0\,mol\,dm^{-3}$ and $1.2 \times 10^{-3}\,mol\,dm^{-3}\,s^{-1}$ when it has fallen to $0.5\,mol\,dm^{-3}$, and so on.

The second advantage of knowing the rate law is that it provides a simple way of predicting the concentration of the reactants (and the products) at any time after the start of the reaction. This is because the rate equation can be solved to find the time dependence of $[A]$. The solution is found in the Appendix, p. 231; here we need only the result. If

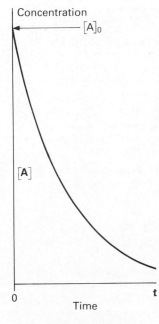

Concentration

$[A]_0$

$[A]$

0   Time t

14.2 The time-dependence of the concentration of a reactant in a first-order reaction the initial concentration of A is $[A]_0$, then at any later time t it will have fallen to $[A]$, where

$$\ln[A] = \ln[A]_0 - k_1 t \quad \text{or} \quad \ln\{[A]/[A]_0\} = -k_1 t. \tag{14.2.2}$$

Figure 14.1 shows a plot of the logarithm of the concentration of a reactant taking part in a first-order reaction. It is a straight line, as the last equation predicts. The time dependence of the concentration itself (not its logarithm) is the *exponential decay* shown in Fig. 14.2. It follows that we can easily test whether a reaction is first-order by checking whether a plot of $\ln[A]$ against time gives a straight line. If it does, then the slope of the line $(-k_1)$ gives the value of the rate coefficient.

As an example of the use of eqn (14.2.2) we can find the time needed for $[A]$ to fall to half its initial value. This is the *half-life*, $t_{\frac{1}{2}}$, of the reaction. When $t = t_{\frac{1}{2}}$ the concentration of A is $\frac{1}{2}[A]_0$. On substituting these values,

$$\ln\{\tfrac{1}{2}[A]_0/[A]_0\} = -k_1 t_{\frac{1}{2}}, \quad \text{or} \quad \ln\tfrac{1}{2} = -k_1 t_{\frac{1}{2}}.$$

Therefore, since $\ln\frac{1}{2} = -\ln 2$, rearrangement gives

$$t_{\frac{1}{2}} = (\ln 2)/k_1 = 0.693/k_1. \tag{14.2.3}$$

For the reaction in which $k_1 = 2.4 \times 10^{-3}\,\text{s}^{-1}$, it follows that the half-life is 289 s, or about 4.8 min.

Note the following very special feature of first-order reactions: eqn (14.2.3) shows that the *half-life is independent of the initial concentration*. In other words, whatever the concentration of reactant present initially, it will fall to half that value in the time $(\ln 2)/k_1$, Fig. 14.3. This is the reason why ^{14}C-dating (Section 1.4) can be used; radioactive decay is a first-order process, and so its half-life is independent of the initial amount of ^{14}C present.

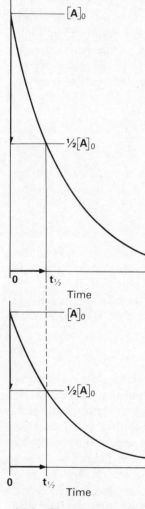

14.3 The half-life of a first-order reaction is independent of the initial concentration

Example: The hydrolysis of sucrose is first-order in the sucrose concentration, and its half-life is 80 min at 20 °C. What proportion of the initial sucrose concentration will remain after (*a*) 160 minutes, (*b*) 240 minutes?

Method: Since the half-life of a first-order reaction is independent of the concentration, each 80-min period results in the halving of the concentration at the start of that period.

Answer: (*a*) 160 minutes is two consecutive half-life periods, and so the initial concentration is halved, and then halved again. Therefore, the initial concentration is reduced to $\frac{1}{4}$ its value.
(*b*) 240 minutes is three successive half-lives, and so the initial concentration is reduced by $\frac{1}{2}$, then $\frac{1}{2}$ again, and then $\frac{1}{2}$ again. Therefore, the concentration is reduced to $\frac{1}{8}$ its initial value.

Comment: This hydrolysis reaction is a first step in the fermentation of alcohol. Brewer's yeast contains the enzyme sucrase which catalyses the reaction.

Principles of physical chemistry

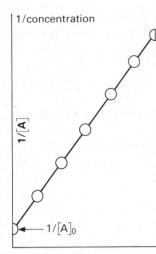

1/concentration

1/[A]

1/[A]₀ (labeled on graph as $1/[A]_0$)

Time

14.4 Investigating a second-order reaction

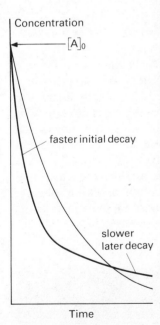

Concentration

$[A]_0$

faster initial decay

slower later decay

Time

14.5 The time-dependence of the concentration of a reactant in a second-order reaction (compared with first-order behaviour)

Second-order reactions

The plot of $\ln[A]$ might be found not to be a straight line. If that turns out to be the case the reaction is not first-order. It is then natural to wonder whether it is *second-order*; that is, a reaction in which the rate is proportional to the square of the concentration of A:

$$rate \propto [A]^2, \quad \text{or} \quad rate = k_2[A]^2. \tag{14.2.4}$$

k_2, the constant of proportionality, is the *second-order rate coefficient*. The equation is the *second-order rate law*, and a reaction it describes is said to follow *second-order kinetics*. Reactions with rate laws of the form $rate \propto [A][B]$, where A and B are species involved in the reaction, are also said to follow second-order kinetics.

Example: The reaction $H_2 + I_2 \rightarrow 2HI$ obeys second-order kinetics. At 400 °C the rate coefficient has the value $2.42 \times 10^{-2}\ dm^3\ mol^{-1}\ s^{-1}$. What is the rate of the reaction when the concentrations of both reactants have reached $0.5\ mol\ dm^{-3}$?

Method: The rate law for the second-order reaction is $rate = k_2[H_2][I_2]$; therefore simply substitute in the stated concentrations.

Answer: The rate of HI production is

$$rate = k_2[H_2][I_2] = (2.42 \times 10^{-2}\ dm^3\ mol^{-1}\ s^{-1})$$
$$\times (0.5\ mol\ dm^{-3}) \times (0.5\ mol\ dm^{-3})$$
$$= 6.05 \times 10^{-3}\ mol\ dm^{-3}\ s^{-1}.$$

Comment: There is no need for the concentrations of the hydrogen and iodine to be the same, because arbitrary amounts of reactants may be introduced initially. The rate law still lets us predict the rate in this more general case, and so it is a succinct summary of all possible reaction rates at 400 °C. Note the dimensions of k_2: 1/(concentration × time). The dimensions of k_1 are 1/(time). In both cases the rate of reaction has the dimensions (concentration)/(time).

One way of testing whether the last equation describes the reaction is to determine the rates at several concentrations, and then to plot them against $[A]^2$; a straight line would confirm second-order kinetics. A much better way is to do the same as for the first-order reaction: solve the rate equation and then plot some simple function of the concentration against the time, hoping for a straight line. The solution of eqn (14.2.4) is described in the Appendix, p. 231. All we need here is the result:

$$\frac{1}{[A]} = \frac{1}{[A]_0} + k_2 t. \tag{14.2.5}$$

We see that *if* the reaction is second-order, a plot of $1/[A]$ against t gives a straight line, and its slope is k_2, Fig. 14.4.

If the investigation confirms that the reaction is second-order, we can use eqn (14.2.5) to predict the concentration of A at any stage. A more useful form is obtained by rearranging it slightly into

$$[A] = \frac{[A]_0}{1 + k_2[A]_0 t}. \tag{14.2.6}$$

The shape of this function is plotted in Fig. 14.5 and compared with the form of a first-order decay.

What is the half-life of a second-order reaction? The time for the concentration of A to fall to $\frac{1}{2}[A]_0$ comes easily from eqn (14.2.5) by setting $t = t_{\frac{1}{2}}$ and $[A] = \frac{1}{2}[A]_0$:

$$k_2 t_{\frac{1}{2}} = \frac{2}{[A]_0} - \frac{1}{[A]_0} = \frac{1}{[A]_0}, \quad \text{or} \quad t_{\frac{1}{2}} = \frac{1}{k_2[A]_0}. \tag{14.2.7}$$

Unlike first-order reactions, the half-life now depends on the initial concentration; the greater the concentration of A the more rapidly it drops to half that value, Fig. 14.6.

Orders of reactions

A reaction might follow kinetics more complicated than either first-order or second-order. If experiment shows that the rate of a reaction involving A and B is of the form

$$\text{rate} = k[A]^x[B]^y, \tag{14.2.8}$$

then it is said to be of *order x* in A, of *order y* in B, and $(x + y)$-*order* overall.

A reaction may be *zeroth-order* in a reactant A, in which case the rate is independent of [A] (because $[A]^0 = 1$, a constant). A disappears at a constant rate until it is all used up. In other words, [A] falls linearly to zero. We shall see an example shortly. In some cases a reactant A may be in such great excess that its concentration remains almost unchanged throughout the reaction. Therefore a second-order rate law of the form $k_2[A][B]$ behaves like a first-order rate law $k_1[B]$ with $k_1 = k_2[A]$, effectively a constant. Such reactions are called *pseudo first-order reactions* and [B] follows a simple exponential decay. They often occur in solution, because if the solvent (such as water) takes part in the reaction, its concentration barely changes.

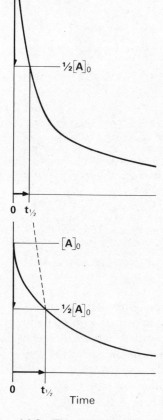

14.6 The half-life of a second-order reaction increases as the initial concentration of reactant is decreased

Example: Account for the following observations on the iodination of propanone in aqueous acid. (*a*) If the propanone concentration is doubled, the rate is doubled. (*b*) If the iodine concentration is doubled, there is no change in the rate. (*c*) If the pH of the medium is reduced from 1.0 to 0.7 the rate doubles.

Method: Find the order of the reaction with respect to each component by noting the power to which the concentration must be raised in order to account for the observations.

Principles of physical chemistry

Answer: (*a*) The first observation shows that the rate is proportional to the concentration of propanone; therefore, since rate $\propto$ [propanone], the reaction is *first-order in propanone*.

(*b*) The rate is independent of the iodine concentration, and so the reaction is *zeroth-order in iodine*.

(*c*) pH = 1.0 corresponds to $[H_3O^+] = 0.10\ mol\ dm^{-3}$; pH = 0.7 corresponds to $[H_3O^+] = 0.20\ mol\ dm^{-3}$. Therefore doubling the hydrogen ion concentration doubles the rate, and so the reaction is *first-order in hydrogen ions*.

Comment: According to this discussion the rate of reaction can be expressed as $k_2[CH_3 \cdot CO \cdot CH_3][H_3O^+]$, and so it is second-order overall. The presence of the hydrogen ion concentration in the rate law indicates an *acid-catalysed reaction*, as we explain shortly.

Quite often reactions cannot be ascribed an order. The apparently simple reaction $H_2 + Br_2 \rightarrow 2HBr$, for instance, has the following complicated rate law

$$\text{rate} = \frac{k'[H_2][Br_2]^{\frac{1}{2}}}{1 + k''([HBr]/[Br_2])}.$$

Although the concept of order breaks down in such cases, the rate law itself remains perfectly well defined: if we want the rate under given conditions, all we need do is feed in the measured values of the rate coefficients and the concentrations of the species involved in the reaction.

14.3 Rates and conditions

Reaction rates depend on the *concentrations* (or pressures) of the reactants, the *temperature*, the presence of *catalysts*, the *state of subdivision* of solids, and, in some cases, the *intensity of light* incident on the system.

The concentration dependence

The rate laws summarize the experimentally observed concentration dependence. In this section we account for it.

In order for molecules to react they must meet. If we imagine molecules moving at random, such as molecules moving chaotically through a gas or floating through a solvent, then the greater their concentrations the more frequently they meet. Hence, the higher the concentration, the greater the rate of reaction.

This accounts very neatly for the rate law for simple gas-phase reactions. Take, for example, the isotope exchange reaction

$$D_2(g) + HCl(g) \rightarrow DH(g) + DCl(g)$$

which experiment shows to obey the overall second-order law

$$\text{rate} = k_2[D_2][HCl]; \qquad k_2 = 0.141\ dm^3\ mol^{-1}\ s^{-1} \quad \text{at } 600\,°C.$$

If we take the view that the reaction occurs as a result of a collision between the D_2 and HCl molecules, then we should expect second-order kinetics. This is because the probability that any D_2 molecule is at a given place is proportional to its concentration, the probability that an HCl molecule is there is proportional to its concentration, and therefore the probability that they are *both* there is proportional to the product of the two concentrations. Hence the rate is expected to be proportional to the product of concentrations, as observed.

There is, however, a very important feature of this style of argument that has to be emphasized: *we cannot predict the form of the rate law simply by looking at the chemical equation.* The rate law has to be determined *experimentally*. After that we can attempt to account for it by proposing a sequence of steps called the *mechanism* of the reaction. Sometimes the mechanism may be very simple; but sometimes it is possible to account for the rate law only in terms of a complicated sequence of steps.

The following example shows what this means and introduces another important point. The chemical equation for the iodination of propanone in acid solution is

$$CH_3 \cdot CO \cdot CH_3(aq) + I_2(aq) \rightarrow CH_2I \cdot CO \cdot CH_3(aq) + HI(aq).$$

It would be quite wrong to assert that the rate equation is rate $= k_2[CH_3 \cdot CO \cdot CH_3][I_2]$ unless we had experimental evidence. In fact, when the rate is investigated experimentally (see the preceding *Example*) it turns out to follow the law

$$rate = k_2[CH_3 \cdot CO \cdot CH_3][H_3O^+].$$

So, although it is second-order overall, the iodine concentration does not affect the rate and so the reaction is zeroth-order in iodine. Although hydrogen ions do not appear as reactants in the chemical equation, they do affect the rate.

The form of the rate law can be explained as follows. Suppose that the reaction proceeds in two steps and that in the first step the propanone is prepared for attack by the iodine, while in the second step the attack occurs. Suppose, too, that the first step is slow and the second fast. Then the overall rate is determined by the rate of the first step, because the second step takes place rapidly once the first has occurred. This is like

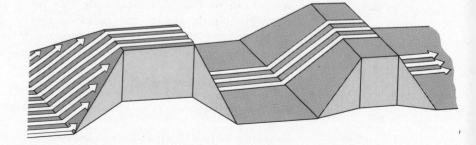

14.7 An analogy for the rate-determining step of a reaction

Principles of physical chemistry

two towns being joined by a good road crossing two bridges; if the first bridge is narrow the traffic will queue to pass over it, but once they are across, the second, broad bridge does not impede them, Fig. 14.7.

When there is a single step that controls the overall rate of a multi-step reaction, it is called the *rate-determining step* of the reaction. In the present example the rate law suggests that the rate-determining step involves a propanone molecule and an H_3O^+ ion. Once the proton has been donated to the propanone the iodine can attack and finish off the reaction quickly. Some enzymes are so efficient that many biological reactions are governed by the rate at which materials can come into contact with them. The rate-determining step for the action of ATP is the rate of supply of materials to triose phosphate isomerase, an exceptionally efficient enzyme.

The rate-determining step in the propanone reaction involves two species, a propanone molecule and a hydrogen ion (H_3O^+). It is therefore called a *bimolecular step*. Sometimes a reaction step consists of a molecule simply shaking itself apart. That is called a *unimolecular step*.

The terms second-order and bimolecular, or first-order and unimolecular should never be confused. The *order* of the reaction is something that must be determined through experiment and inspection of the rate law; the *molecularity* is the number of species involved in a particular step in the reaction scheme. The observed rate law, even if it is second-order, may be the outcome of a complex chain of unimolecular and bimolecular processes that has to be unravelled by detailed detective work.

The temperature dependence

Most reactions go faster at higher temperatures. When Arrhenius studied the effect of temperature on chemical reactions (in about 1889) he found that the temperature dependence of the rate coefficient could usually be expressed as

$$\text{Arrhenius law: } k = Ae^{-E_a/RT} \qquad (14.3.1)$$

where A and E_a depend on the reaction. E_a is the *activation energy* and A is the *pre-exponential factor*. The Arrhenius law applies to all kinds of reactions. Even luminescent tropical butterflies have been found to flash more rapidly on warmer nights in accord with it. The activation energies of some reactions are given in Table 14.1.

Table 14.1   Activation energies of some reactions, $E_a/\text{kJ mol}^{-1}$

Reaction	Conditions	$E_a/\text{kJ mol}^{-1}$
$C_2H_4 + H_2 \rightarrow C_2H_6$	gas phase	180
$2HI \rightarrow H_2 + I_2$	gas phase	184
	Pt catalyst	59
$H_2 + Cl_2 \rightarrow 2HCl$	photochemical	25
$CH_3Cl + CH_3O^- \rightarrow CH_3OCH_3 + Cl^-$	methanol solvent	100

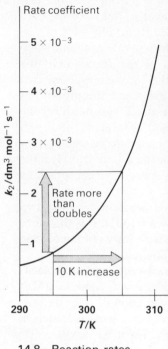

Rate coefficient

5×10^{-3}

4×10^{-3}

3×10^{-3}

$k_2/\mathrm{dm^3\,mol^{-1}\,s^{-1}}$

2

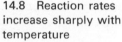

Rate more than doubles

1

10 K increase

290   300   310

T/K

14.8 Reaction rates increase sharply with temperature

Example: Find the ratio of the rate coefficients at 20 °C and 30 °C when the activation energy of a reaction is 50 kJ mol⁻¹.

Method: Use the Arrhenius expression, eqn (14.3.1), with the temperatures expressed in kelvins. The A-factors cancel when ratios are taken. Evaluate E_a/RT at each temperature.

Answer: At 30 °C,

$$E_a/RT = \frac{(50 \times 10^3\,\mathrm{J\,mol^{-1}})}{(8.314\,\mathrm{J\,K^{-1}\,mol^{-1}}) \times (303.15\,\mathrm{K})} = 19.84.$$

At 20 °C,

$$E_a/RT = \frac{(50 \times 10^3\,\mathrm{J\,mol^{-1}})}{(8.314\,\mathrm{J\,K^{-1}\,mol^{-1}}) \times (293.15\,\mathrm{K})} = 20.51.$$

The ratio of rate coefficients is therefore

$$\frac{k(30\,°\mathrm{C})}{k(20\,°\mathrm{C})} = \frac{e^{-19.84}}{e^{-20.51}} = \frac{2.42 \times 10^{-9}}{1.24 \times 10^{-9}} = 1.95.$$

Comment: This calculation is the basis of the common remark that a 10 K rise in temperature doubles the rates of reactions, Fig. 14.8. Many reactions do have activation energies of around 50 kJ mol⁻¹, and for them this prediction is valid. On the other hand, many do not, and the effect of temperature is quite different. It is always easy to predict the effect of temperature once the activation energy is known, simply by repeating this calculation with the appropriate value of E_a.

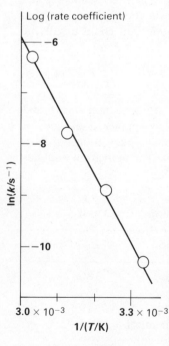

Log (rate coefficient)

-6

-8

$\ln(k/\mathrm{s^{-1}})$

-10

3.0×10^{-3} 3.3×10^{-3}

$1/(T/\mathrm{K})$

14.9 An Arrhenius plot

Activation energies can be found by measuring reaction rates at different temperatures. First the Arrhenius expression is rearranged by taking the logarithm of both sides:

$$\ln k = \ln A - E_a/RT. \qquad (14.3.2)$$

Then the rate coefficient is measured at several temperatures and $\ln k$ is plotted against $1/T$. This should give a straight line of slope $-E_a/R$, Fig. 14.9. This procedure is called an *Arrhenius plot*. (The intercept of the plot with the $\ln k$ axis at $1/T = 0$ is equal to $\ln A$, and so A may also be obtained from the same set of experiments.)

Example: The decomposition of dinitrogen pentoxide (N_2O_5) was followed at different temperatures and the rate coefficient for the reaction was found to vary as follows:

T/K	300	310	320	330
$k_1/\mathrm{s^{-1}}$	3.38×10^{-5}	1.35×10^{-4}	4.98×10^{-4}	1.83×10^{-3}

Find the activation energy of the reaction.

Principles of physical chemistry

Method: Construct an Arrhenius plot by drawing up a table of $\ln(k_1/s^{-1})$ and $1/(T/K)$. The slope of the graph is $-E_a/R$.

Answer:

$1/(T/K)$	3.33×10^{-3}	3.23×10^{-3}	3.13×10^{-3}	3.03×10^{-3}
$\ln(k_1/s^{-1})$	-10.3	-8.91	-7.60	-6.30

These points are plotted in Fig. 14.9. The slope is

$$\text{slope} = \frac{(8.12 - 9.45)}{(0.10 \times 10^{-3})} = -1.33 \times 10^{4}.$$

It follows that $-E_a/R = (-1.33 \times 10^{4})$ K, or $E_a = (1.33 \times 10^{4} \text{ K}) \times R = (1.33 \times 10^{4} \text{ K}) \times (8.314 \text{ J K}^{-1} \text{ mol}^{-1}) = 111 \text{ kJ mol}^{-1}$.

Comment: Pay special attention to the way that the slopes of graphs are handled (see p. 239).

The reason why reactions have the Arrhenius temperature dependence is quite simple. Not only must species meet in order to undergo reaction, but they must also have enough energy to break and reorganize their bonds. In other words, in the case of a gas-phase bimolecular step, not only must the species collide, but they must smash together with sufficient energy for atomic rearrangement to occur.

The key to explaining the actual form of the Arrhenius rate law is the Maxwell-Boltzmann distribution of molecular speeds (Section 4.2). Figure 14.10 reproduces the main pieces of information. As the temperature is increased, the proportion of particles having high speeds, and therefore high kinetic energies, increases. The molecules having kinetic energies of at least E_a are those corresponding to the shaded areas in the illustration; it is clear that many more have at least that energy at high temperatures than at low. The mathematical form of the dependence on the temperature is

$$\left.\begin{array}{c}\text{Proportion of molecules having at least}\\ \text{the energy } E_a \text{ at the temperature } T\end{array}\right\} = e^{-E_a/RT}.$$

This is exactly what is required to account for the Arrhenius law.

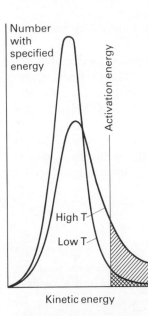

14.10 The proportions of molecules having at least the activation energy (shaded regions) at two temperatures

Number with specified energy

Activation energy

High T

Low T

Kinetic energy

14.4 Catalysis

A *catalyst* is a substance that increases the rate of a chemical reaction without itself being consumed. It does not affect the equilibrium position of the reaction, but it does increase the rate at which equilibrium is reached.

The action of catalysts

A catalyst is more an active bystander than a mere onlooker. Although

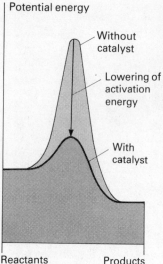

14.11 The effect of a catalyst on the activation energy

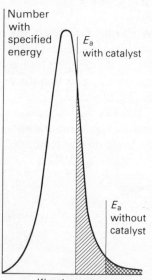

14.12 The proportion of molecules having at least the activation energy (shaded regions) with and without a catalyst (at the same temperature)

the catalyst does not occur in the reaction equation, as in the Haber-Bosch synthesis of ammonia, $N_2(g) + 3H_2(g) \rightleftharpoons 2NH_3(g)$, the fact that it affects the reaction rate must mean that it is involved in intermediate steps.

A catalyst works by providing an alternative reaction pathway with a lower activation energy. This is illustrated in Fig. 14.11. For a given temperature a higher proportion of reactant molecules can undergo reaction when the catalyst is present (small E_a) than when it is not (large E_a), Fig. 14.12.

The operation of catalysts depends both on their nature and the reaction, but they have several features in common.

1 A catalyst increases the rates of both the backward and forward reactions, and does not affect the position of equilibrium.

2 While it is often the case that very small amounts of catalyst are sufficient, some need to be present in large quantities (as for aluminium chloride in Friedel-Crafts acylation; e.g., $C_6H_6(l) + CH_3COCl(l) \rightarrow C_6H_5COCH_3(l) + HCl(g)$).

3 Sometimes a *promoter* is necessary. This is a substance that improves the performance of the catalyst itself. For instance, trace quantities of potassium and aluminium oxides are added to the finely divided iron catalyst used in the main stage of the Haber-Bosch process.

4 Although a catalyst does not undergo a net chemical change, its involvement in the reaction may result in it changing its physical form. For instance, it might crumble. An example is provided by the oxidation of ammonia to nitrogen oxide (NO), known as the *Ostwald process*, which is carried out over a platinum/rhodium gauze that has to be changed periodically on account of the surface roughening induced by the reaction.

5 Catalysts may be exceptionally specific, affecting one reaction strongly but leaving a closely similar reaction almost unaffected. *Enzymes*, naturally occurring biologically active catalysts, are very specific. For instance, urease catalyses the hydrolysis of carbamide (urea, NH_2CONH_2) but has no effect on ethanamide (CH_3CONH_2).

6 Catalysts can be *poisoned* (made ineffective). For instance, carbon monoxide poisons platinum used as a catalyst in the synthesis of water. Catalyst poisons are often animal poisons too, and operate by blocking enzyme controlled reactions.

The classification of catalysts

The two main classes of catalysts are homogeneous catalysts and heterogeneous catalysts. A *homogeneous catalyst* is in the same phase as the reactants. For instance, a homogeneous catalyst for the gas-phase reaction $2SO_2(g) + O_2(g) \rightleftharpoons 2SO_3(g)$ is the gas nitrogen dioxide, NO_2. A *heterogeneous catalyst*, on the other hand, is in a different phase from the reactants. This normally means that the catalyst is a solid while the reactants are liquids or gases. The catalysts used in the Haber-Bosch process are all heterogeneous. Table 14.2 lists some of the heterogeneous catalysts now available.

Principles of physical chemistry

Table 14.2 Heterogeneous catalysts

Class	Function	Examples
Metals	hydrogenation, dehydrogenation	Fe(Haber), Ni, Pd, Pt
Semiconducting oxides	oxidation, dehydrogenation	V_2O_5 (contact process)
Insulating oxides	dehydration	Al_2O_3, SiO_2
Acids	polymerization, isomerization, cracking	H_2SO_4, SiO_2

How homogeneous catalysts work

Homogeneous catalysis is normally the result of the formation of some intermediate compound which is more easily attacked than is the original reactant itself. Thus in the reaction $A + B \rightarrow$ products, a catalyst C may combine first with A to form AC. That species is susceptible to attack by B, resulting in the regeneration of C and the formation of products. If this overall process has a lower activation energy than has the original single-step reaction it will proceed more quickly.

Important examples of homogeneous catalysis are *acid catalysis* and *base catalysis*. Many organic reactions proceed by one or other mechanism, and sometimes by both. In acid catalysis the important step is the transfer of a proton to one of the reactants, so forming AH^+. The protonated species is then attacked. This occurs in the acid hydrolysis of esters and, as we saw, in the iodination of propanone. A base-catalysed reaction is the opposite, one of the reactants transferring a proton to a base that has been added. The resulting species A^- is then attacked. This is the basis of the halogenation of some organic compounds.

How heterogeneous catalysts work

A heterogeneous catalyst is active on account of the processes that occur at its surface. The first step is the *adsorption* of a reactant on to the surface. (*Ad*sorption, attachment to a surface, must be distinguished from *ab*sorption, penetration into the bulk.) Adsorption often takes place by the species breaking apart as it bonds to the surface, Fig. 14.13(a). This makes it open to attack by other species already adsorbed on the surface, and free enough to migrate over it, or by collisions with species in the gas or liquid above the surface, Fig. 14.13(b).

As an example, the key steps in the oxidation of propene are illustrated in Fig. 14.14. The reaction proceeds at a bismuth molybdate surface, and involves some very complicated physical chemistry. The first step, Fig. 14.14(a), is the adsorption of the molecule by loss of a hydrogen atom and the formation of a carbon-surface bond. The hydrogen escapes with a surface oxygen atom, Fig. 14.14(b), and goes on to form water. The

(a)

(b)

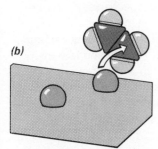

14.13 Heterogeneous catalysis (a) adsorption and dissociation, (b) attack

14.14 The oxidation of propene. (*a*) Approach and adsorption of $CH_2 : CH \cdot CH_3$; (*b*) loss of H and its removal of a surface O; (*c*) loss of second H; (*d*) extraction of surface O by adsorbed hydrocarbon; (*e*) escape of propenal and regeneration of surface by O_2

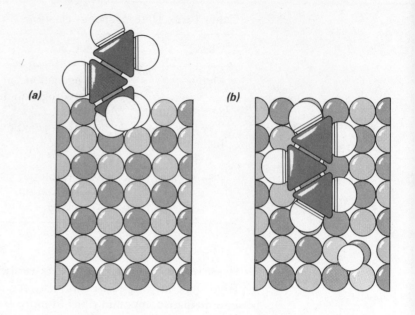

(a)   (b)

hydrocarbon molecule loses another hydrogen atom, Fig. 14.14(*c*), drags another oxygen atom out of the surface, Fig. 14.14(*d*), and escapes as propenal, $CH_2 : CHCHO$, which can then be used as the starting point for other synthetic processes. The removal of the oxygen atoms leaves gaps in the surface which are filled by attack by molecular oxygen from the gas, Fig. 14.14(*e*).

Although the last step reforms the catalyst it is clear that its surface has to survive violent changes, and may easily break up under the stress. That is one of the reasons why a catalyst may need to be *supported* on some medium in order to increase its surface strength. Supporting a catalyst has the additional advantage of increasing its surface area. An example is phosphoric acid, which is supported on a material based on silica (Celite) when it is used to catalyse the direct hydration of ethene to ethanol. Other useful supports are aluminium oxide and charcoal.

It is not hard to see that a large amount of physical chemistry is involved in the thorough investigation of the mechanisms of both homogeneous and heterogeneous catalysis. Every step is open to investigation using the techniques we have touched on in the preceding chapters. The structures of solid catalysts can be studied with X-ray and low-energy electron diffraction. The species involved in each step of an involved reaction sequence can be identified spectroscopically. The positions of equilibrium of reactions can be investigated using the techniques of thermodynamics.

All this is of vital importance. The whole of everyday life depends in some degree on catalysis, either because it is essential to the petrochemical, polymer, and fertilizer industries or because it is being used to control pollution and increase the efficiency of our deployment of energy. Catalysts also regulate all the processes of biology, from the growth of a leaf to the mental processes involved in thinking and learning.

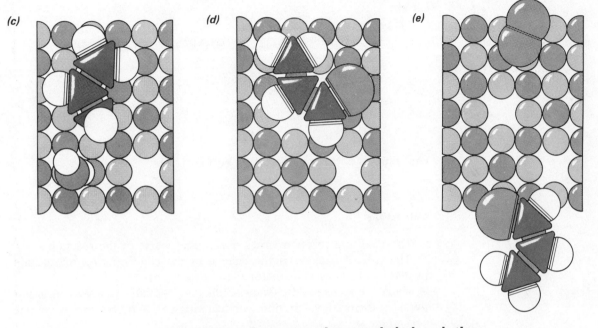
(c)   *(d)*   *(e)*

Appendix: Rate equations and their solutions

The concentration of a reactant A changes by the infinitesimal quantity d[A] in the infinitesimal interval dt. It follows that the *instantaneous rate* of a reaction at some time, Fig. 14.15, is

$$\text{rate} = -d[A]/dt.$$

(The negative sign is introduced so that the rate is positive when the concentration of the reactant decreases with time, d[A] then being negative.) The precise form of the first-order rate equation, eqn (14.2.1), is therefore

$$-d[A]/dt = k_1[A].$$

This is a standard form of differential equation, with the solution

$$[A] = a e^{-k_1 t}, \quad \text{with } a \text{ a constant.}$$

Check:

$$\frac{d[A]}{dt} = a \frac{d}{dt} e^{-k_1 t} = -k_1 a e^{-k_1 t} = -k_1[A], \quad \text{as required.}$$

At $t = 0$, $[A]_0 = a e^0 = a$, since $e^0 = 1$. Therefore $a = [A]_0$. Taking logs of the solution leads to eqn (14.2.2) directly.

In the case of second-order kinetics, eqn (14.2.4) becomes

$$-d[A]/dt = k_2[A]^2.$$

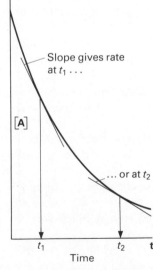

Concentration

Slope gives rate at t_1 ...

[A]

... or at t_2

t_1 t_2 t

Time

14.15 The reaction rate and the slope of a concentration/time graph

Try a solution of the form

$$[A] = \frac{[A]_0}{1 + k_2[A]_0 t} = [A]_0\{1 + k_2[A]_0 t\}^{-1}.$$

Check:

$$\frac{d[A]}{dt} = -[A]_0\{1 + k_2[A]_0 t\}^{-2} \frac{d}{dt}(k_2[A]_0 t) = -\frac{k_2[A]_0^2}{(1 + k_2[A]_0 t)^2}$$

$$= -k_2\left\{\frac{[A]_0}{1 + k_2[A]_0 t}\right\}^2 = -k_2[A]^2 \quad \text{as required;}$$

also

$$[A] = [A]_0 \quad \text{at} \quad t = 0.$$

The solution is the one quoted in eqn (14.2.6).

Summary

1   The rate of a reaction varies during the course of the reaction.

2   The *rate of reaction* is the change of the concentration of reactant divided by the time it takes for the change to occur.

3   Rates are *measured* by *following the time dependence of the concentration* by spectroscopy, titration, conductivity, pH, and pressure or volume measurements.

4   In order to measure a concentration the reaction must be *quenched* (slowed to an imperceptible rate).

5   A *first-order rate equation* is rate $= k_1[A]$. k_1 is the *rate coefficient*. The time dependence of the concentration is given by eqn (14.2.2).

6   The *half-life* of a reaction is the time taken for the concentration to decrease to one half its initial value.

7   For a first-order reaction the half-life is independent of the initial concentration and is related to the rate coefficient by $t_{\frac{1}{2}} = (\ln 2)/k_1$.

8   A *second-order rate equation* is rate $= k_2[A]^2$. k_2 is the rate coefficient. The time dependence of the concentration is given by eqn (14.2.5).

9   For a second-order reaction the half-life decreases with increasing initial concentration and is given by $t_{\frac{1}{2}} = 1/k_2[A]_0$.

10   If the rate of a reaction between A and B has the form rate $= k[A]^x[B]^y$ then the *order in* A is x, the order in B is y, and the *overall order* is $(x + y)$.

11   The rate equation *cannot be predicted* from the chemical equation. It must be determined experimentally.

12   Reactions often occur in several steps. If one step is much slower than the others it is termed the *rate-determining step*.

13   A rate equation is explained by proposing a *mechanism* for the reaction, a sequence of individual reaction steps.

14   The *molecularity* is the number of species involved in a given step in the reaction mechanism.

15   The temperature dependence of rate coefficients usually follows the

Arrhenius law: $k = Ae^{-E_a/RT}$. E_a is the *activation energy*, A is the *pre-exponential factor.*

16 An *Arrhenius plot* of ln k against $1/T$ has a slope $-E_a/R$, and is used to measure the activation energy.

17 The rates of many reactions typically double for a temperature rise of $10\,°C$ near room temperature.

18 The Arrhenius law can be explained on the basis that molecules (*a*) must meet and (*b*) must have at least the energy E_a in order to react.

19 A *catalyst* increases the rate of a chemical reaction without itself being consumed.

20 A catalyst (*a*) does not affect the position of equilibrium; (*b*) is effective in small quantities; (*c*) may be physically altered; (*d*) is often specific; (*e*) may be promoted or poisoned; (*f*) may require a support.

21 Catalysts are either in the same phase as the reactants (*homogeneous*) or in a different phase (*heterogeneous*).

22 Important examples of homogeneous catalysis are *acid catalysis* and *base catalysis*. Frequently an *intermediate compound* is formed.

23 Heterogeneous catalysis occurs by *adsorption* of one or both reactants on to the catalyst surface followed by reaction and then *desorption*.

Problems

1 Define the terms *rate, rate coefficient, order of reaction* and *half-life.*

2 What is the difference between the *order* and the *molecularity* of a reaction?

3 Explain how a reaction may be tested for (*a*) first-order kinetics, (*b*) second-order kinetics.

4 The rate coefficient of a first-order reaction was measured as $1.11 \times 10^{-3}\,s^{-1}$. What is the half-life of the reaction? What time is needed for the initial concentration to fall to (*a*) $\frac{1}{8}$ its initial value, (*b*) $\frac{3}{4}$ its initial value?

5 What fraction of the initial concentration in the reaction in the last Question will remain after (*a*) $15\,s$, (*b*) $1\,h$?

6 The rate of decomposition of hydrogen peroxide in aqueous solution was measured by titration against aqueous potassium manganate(VII) solution. Equal volumes of solution were withdrawn at various times, and titrated:

t/min	0	10	20
$V(KMnO_4)$/cm^3	45.6	27.6	16.5

Confirm that the reaction is first-order in the peroxide and find the half-life and rate coefficient.

7 In an experiment to determine the rate of inversion of sugar (*i.e.*, the rate of its conversion into glucose and fructose) the following data were obtained using a polarimeter:

t/min	0	10	20	30	40	80	∞
Angle/deg	64.8	57.6	51.0	44.8	39.2	20.6	−28.2

Plot the data and determine the order of the reaction with respect to sucrose. Find the rate coefficient and calculate the half-life of the reaction.

8 Explain why rates of reaction usually increase with temperature, and define and explain the term *activation energy.*

9 What do you understand by the term *order* when applied to a chemical reaction?

Given data on the variation of concentration with time, give two methods which may be used to determine the order of that reaction.

Propanone (acetone) reacts with iodine in the presence of acid according to the equation

$$CH_3COCH_3 + I_2 \rightarrow CH_3COCH_2I + H^+ + I^-$$

In an experiment, 5 cm³ of propanone, 10 cm³ of sulphuric acid of concentration 1.0 mol dm⁻³ and 10 cm³ of a solution of iodine of concentration 1.0×10^{-3} mol dm⁻³ were mixed, made up to 100 cm³ with distilled water and placed in a thermostat. Every five minutes samples were removed and titrated against a solution of sodium thiosulphate. The experiment was repeated but using 20 cm³ of sulphuric acid in the reaction mixture. The following results were obtained.

Time/min		5	10	15	20	25
Titre/cm³	for 10 cm³ acid	18.5	17.0	15.5	14.0	12.5
	for 20 cm³ acid	17.0	14.0	11.0	8.0	5.0

From plots of titre against time, determine the order of reaction with respect to iodine and with respect to [H⁺].

How would you determine the order with respect to propanone?

(*Oxford and Cambridge*)

10   The rate of a reaction between A and B varies as follows:

rate = $k[A]^2[B]$

(a) What is the overall order of the reaction?

(b) If the concentration of only one of the reactants can be doubled, which would give the greater increase in the overall rate?

(c) Give the reasons for your answer to (b).

(d) Suggest two other ways by which the rate of the reaction might be increased.

(*Based on JMB*)

11   (a) Explain what is meant by the terms (i) *velocity constant* (or *rate constant*), (ii) *activation energy*.

(b) Nitrogen dioxide decomposes in the gaseous phase into nitrogen oxide and oxygen. Suggest an experimental method for determining the rate of this reaction.

(c) The following results were obtained for the velocity constant k of this reaction:

k/dm³ mol⁻¹ s⁻¹	*Temperature*/K
3.16	650
28.2	730
158	800
1120	900
5010	1000

Use this information to find the activation energy for this reaction.

(*Oxford S*)

12   Ethanoic acid can be prepared by reacting methanol in the liquid state with

carbon monoxide using a *homogeneous liquid-phase catalyst* – a complex compound of rhodium with iodine as a promoter.

$$CH_3OH + CO \rightarrow CH_3\!-\!\underset{\underset{O}{\|}}{C}\!-\!OH$$

The reaction is of *zero order with respect to both reactants* – methanol and carbon monoxide. With respect to the catalyst components, *the reaction rate is directly proportional to both the rhodium complex catalyst and an intermediate, iodomethane*, CH_3I, formed during the reaction.

        (*i*) Explain what is meant by the following phrases
            (1) *homogeneous liquid-phase catalyst*
            (2) *the reaction is of zero-order with respect to both reactants.*
       (*ii*) Write a rate equation showing the dependence of the rate of reaction on the concentrations of each of the rhodium complex and iodomethane.
      (*iii*) State the overall order of reaction.

(AEB 1979)

13   When ethanal is heated to about 800 K, it decomposes by a *second-order reaction* according to the equation

$$CH_3CHO(g) \rightarrow CH_4(g) + CO(g)$$

If, however, the ethanal is heated to the same temperature with iodine vapour, the iodine *catalyses* the reaction; this latter reaction is *first-order with respect to both iodine and ethanal* and has a lower *activation energy* than the reaction with ethanal alone.

     (*a*) Explain the meaning of the terms in italics.
     (*b*) Suggest a set of experiments by which you could confirm that the uncatalyzed reaction is second-order.
     (*c*) Indicate briefly what experiments would be necessary to show that the reaction with iodine present has a lower activation energy than the decomposition of ethanal itself.

(London)

14   Define the order and rate constant for a chemical reaction. The hydrolysis of sucrose is found to be first-order with respect to sucrose and first-order with respect to hydrogen ion concentration, with a rate constant $k = 2.3\,dm^3\,mol^{-1}\,s^{-1}$. Calculate the time for the sucrose concentration to fall to $0.05\,mol\,dm^{-3}$ in solutions with the following initial concentrations: (*i*) $0.1\,mol\,dm^{-3}$ sucrose, $0.1\,mol\,dm^{-3}$ HCl; (*ii*) $0.1\,mol\,dm^{-3}$ sucrose in a buffer solution with pH $= 4$.

(Oxford Entrance)

15   What do you understand by the terms *rate expression* and *reaction order*? Suggest one method each for measuring the rates of the following reactions:
     (*a*) The hydrolysis of ethyl ethanoate (ethyl acetate).
     (*b*) The oxidation of hydriodic acid by hydrogen peroxide.
The equilibrium between two isomers, A and B can be represented:

$$A \underset{k_2}{\overset{k_1}{\rightleftharpoons}} B$$

where k_1 and k_2 are rate constants for the forward and reverse reactions, respectively. Starting with a non-equilibrium mixture of concentrations $[A]_0 = a$ and $[B]_0 = b$ it was found that x moles of A had reacted after time t. Give an expression for the rate, dx/dt, and hence show that the integrated rate expression is

$$\ln\left(\frac{p}{p-x}\right) = (k_1 + k_2)t$$

where

$$p = \frac{k_1 a - k_2 b}{k_1 + k_2}.$$

After 69.3 minutes $x = p/2$; calculate k_1 and k_2 if the equilibrium constant $K = 4$.

(Cambridge Entrance)

16   What do you understand by the *rate constant* of a chemical reaction? Explain the concept of *activation energy* and show how it may be used to explain why some reaction rates increase dramatically for only a modest increase in temperature.

A reaction was found to have the following temperature dependence of the rate constant:

T/K	500	555	625	714
$k/\mathrm{dm^3\,mol^{-1}\,s^{-1}}$	0.025	0.36	5.5	79

Derive the activation energy for this reaction by plotting a suitable graph. Does this graph provide any other information about the reaction?

(Oxford Entrance)

Background information: quantities and equations

1 Symbols

We have used internationally accepted symbols and subscripts. The states of substances are denoted g(gas), l (liquid), s (solid), aq (aqueous solution). Standard states are indicated by the superscript $^{\ominus}$. Molar quantities, of which more below, are labelled with a subscript m. The sign $\approx$ means 'approximately equal to'. $x > y$ means x greater than y; $x < y$ means x less than y. ln x is the natural logarithm (base e) of x, and lg x is the logarithm to the base 10.

2 Units

We have used the international system of units ('SI') based on the kilogram (kg), metre (m), second (s), ampere (A), kelvin (K) and mole (mol). The prefixes used are as follows:

p (pico)	n (nano)	μ (micro)	m (milli)
10^{-12}	10^{-9}	10^{-6}	10^{-3}

c (centi)	d (deci)	k (kilo)	M (mega)
10^{-2}	10^{-1}	10^3	10^6

The units of quantities include the following:

energy: joule, $1\,J = 1\,kg\,m^2\,s^{-2}$ power: watt, $1\,W = 1\,J\,s^{-1}$

force: newton, $1\,N = 1\,kg\,m\,s^{-2}$ pressure: pascal, $1\,Pa = 1\,N\,m^{-2}$

charge: coulomb, $1\,C = 1\,A\,s$ potential difference: volt,
$$1\,V = 1\,J\,A^{-1}\,s^{-1} = 1\,J\,C^{-1}$$

In numerical calculations we include the units and cancel and multiply them just like ordinary algebraic symbols. This is a good way of avoiding mistakes because if the units turn out to be wrong the calculation is certainly wrong (but the opposite is not necessarily true!) An example of this *quantity calculus* is the calculation of the value of R in Section 4.1.

3 Amounts of substance

The clumsy name 'amount of substance' has a specific meaning, and is sharper than the colloquial term 'amount'. Its unit is the *mole*, mol; 1 mol means a specified number of objects. That number is the number of atoms in exactly 12 g of ^{12}C. The best count to date gives the number as 6.022×10^{23}. Therefore, whenever we have that number of objects (atoms, molecules, ions, photons, horses ...) we can report that the amount of substance is 1 mol. Note, however, that we have to specify the objects. '1 mol of water' is meaningless (molecules of water? glasses of water? ...); we should say 1 mol H_2O or even 1 mol H_2O molecules. The *number of objects per unit amount of substance* is denoted L and is called *Avogadro's constant*. The best modern value is $L = 6.022 \times 10^{23} \, mol^{-1}$. Note that L is not a number, it is a *constant* (it has units). The symbol for amount of substance is n (*not* n mol, any more than m grams). If the amount of substance in a sample of sodium chloride is 2.3 mol NaCl (for instance), we can write $n = 2.3$ mol NaCl and know that the number of NaCl units present is $nL = (2.3 \, mol) \times (6.022 \times 10^{23} \, mol^{-1}) = 1.4 \times 10^{24}$.

4   Concentration, molarity, and molality

Concentration is the *amount of substance per unit volume*. If the amount of substance is expressed in moles and the volume of the solution expressed in cubic decimetres (litres, $1 \, l = 1 \, dm^3$), the concentration is in $mol \, dm^{-3}$. *Molarity* is a synonym for concentration, but its use is now frowned upon; the statement 'a 2.0 M solution' is still sometimes encountered, but the modern expression would be 'a solution of concentration $2.0 \, mol \, dm^{-3}$'. *Molality* is the amount of substance of solute per unit mass of solvent. It is normally expressed as so many $mol \, kg^{-1}$, meaning so many moles of solute per kilogram of solvent.

5   Molar quantities

The molar mass, M_m, is the mass per unit amount of substance. It is normally expressed as so many grams per mole ($g \, mol^{-1}$). It has the same numerical value as M_r; that is, $M_m = M_r \, g \, mol^{-1}$. If the mass of a sample is M, the amount of substance present is $n = M/M_m$. In general a molar quantity (*e.g.*, molar entropy) is the quantity per unit amount of substance. Therefore if a sample contains an amount of substance n and its entropy is S, the molar entropy is $S_m = S/n$. In the text we encounter molar enthalpy, entropy, Gibbs function, and heat capacity. The molar conductivity is the odd man out; it is conductivity divided by concentration.

6   Chemical equations

The arrow $\rightarrow$ is used to denote a definite direction of reaction; the double half-arrow $\rightleftharpoons$ is used to denote an equilibrium. We refer to the substances on the left of a chemical equation as the 'reactants', and those on the right as the 'products' even though at equilibrium the terms lose their distinction. We have normally specified the phase of the reactants and the products, as in $2H_2(g) + O_2(g) \rightarrow 2H_2O(l)$. The interpretation of the reaction is as follows. *Either* we refer to individual molecules, when it states that two molecules of hydrogen gas react with one molecule of oxygen gas to produce two molecules of liquid water. *Or* we refer to amounts of substance. Then we say that 2 mol H_2 molecules (in the gas phase) react with 1 mol O_2 molecules (in the gas phase) to produce 2 mol H_2O molecules (as liquid). It is an elementary but necessary requirement that the equation should *balance*, that is, the number of atoms of each element must be the same on both sides of the equation. Furthermore, if charged species are involved, the net charge must be the same on each side of the reaction.

7 Tables and graphs

All tables have pure numbers for their entries. This is achieved by dividing the quantity at the head of the column (*e.g.*, the molar ionization energy I_1 in Table 1.2) by its units. Thus $I_1/kJ\,mol^{-1}$ is a dimensionless number because the units of I_1 are cancelled. Since Table 1.2 shows that $I_1/kJ\ mol^{-1} = 1312$ for hydrogen, the molar ionization energy itself is $I_1 = 1312\,kJ\,mol^{-1}$. Graphs are drawn with their axes labelled with pure numbers (obtained in the same way). The slope of a graph is therefore also a dimensionless number.

Further reading

Chapter 1 *Quanta: a handbook of concepts*, P. W. Atkins; Clarendon Press, Oxford, 1974.
Manhattan project, S. Groueff; Collins, 1967 (the making of the atomic bomb).
Chemistry: principles and applications, G. T. Miller; Wadsworth, 1976, chs 2 and 3.
'Synthetic elements IV', G. T. Seaborg and J. L. Bloom; *Scientific American*, **220,** 56 (April 1969).
Chemistry, man and society, M. M. Jones *et al.*; W. B. Saunders, 1976, chs 3 and 6–8.
The periodic table of the elements, R. J. Puddephatt; Clarendon Press, Oxford, 1972.

Chapter 2 *Physical chemistry*, P. W. Atkins; Oxford University Press and W. H. Freeman, 2nd edn, 1982, ch. 15.
The shape and structure of molecules, C. A. Coulson; Oxford University Press, 1973.
The chemical bond, L. C. Pauling; Cornell University Press, 1967.
Chemical systems, J. A. Campbell; W. H. Freeman, 1970, ch. 2.
Valency and molecular structure, E. Cartmell and G. W. A. Fowles; Butterworths, 4th edn, 1977.
General chemistry, readings from *Scientific American*, chs 5 and 25 (computer calculations; aromatic molecules).
Introducing chemistry, H. Rossotti; Penguin, 1975, part II.

Chapter 3 *Spectroscopy*, D. H. Whiffen; Longman, 1966.
The determination of molecular structure, P. J. Wheatley; Oxford University Press, 2nd edn, 1968, chs 6 and 7.
Physical methods and molecular structure, Open University Press, 1977, chs 1 and 2.
'Laser spectroscopy', M. S. Feld and V. S. Letokhov; *Scientific American*, **229,** 69 (Dec. 1973).
'The search for life on Mars', N. H. Horowitz; *Scientific American*, **237,** 52 (Nov. 1977).
'The causes of color', K. Nassau; *Scientific American*, **243,** 106 (October 1980).

Chapter 4 *Physical Chemistry*, P. W. Atkins; Oxford University Press and W. H. Freeman, 2nd edn, 1982, chs 1 and 24.
Asimov's biographical encyclopedia of science and technology, I. Asimov; Pan, 1975.
Chemical systems, J. A. Campbell; W. H. Freeman, 1970, ch. 6.

Chapter 5 *The third dimension in chemistry*, A. F. Wells; Clarendon Press, Oxford, 1956.
Inorganic chemistry, J. E. Huheey; Harper and Row, 1975, ch. 3.
'Materials', *Scientific American*, **217,** (September 1967).

Principles of physical chemistry

The hydrogen bond, J. C. Speakman; Royal Institute of Chemistry, 1975.
Seven solid states, W. J. Moore; Benjamin, 1967.

Chapter 6   *Physical chemistry*, P. W. Atkins; Oxford University Press and W. H. Freeman, 2nd edn, 1982, ch. 7.
General chemistry, readings from *Scientific American*, chs 10 and 12 (molecular motion; ice).
'Liquid crystal display devices', G. H. Heilmeier; *Scientific American*, **222,** 100 (April 1970).

Chapter 7   *Physical chemistry*, P. W. Atkins; Oxford University Press and W. H. Freeman, 2nd edn, 1982, ch. 8.
Elements of physical chemistry, S. Glasstone and D. Lewis; Macmillan, 2nd edn, 1960, ch. 11.
Metals in the service of man, W. Alexander and A. Street; Penguin, 6th edn, 1976, chs 5 and 11.
Thermodynamics 1 and 2, Open University Press, 1975.

Chapter 8   *Physical chemistry*, P. W. Atkins; Oxford University Press and W. H. Freeman, 2nd edn, 1982, ch. 26.
Experimental approach to electrochemistry, N. J. Selley; Edward Arnold, 1977.

Chapter 9   *Physical chemistry*, P. W. Atkins; Oxford University Press and W. H. Freeman, 2nd edn, 1982, chs 2–4.
Bioenergetics, A. L. Lehninger; Benjamin, 1965.
Some thermodynamic aspects of inorganic chemistry, D. A. Johnson; Cambridge University Press, 1968.
Tables of physical and chemical constants, G. W. C. Kaye and T. H. Laby; Longman, 14th edn, 1973.

Chapter 10   *Physical chemistry*, P. W. Atkins; Oxford University Press and W. H. Freeman, 2nd edn, 1982, ch. 9.
Chemistry: principles and applications, G. T. Miller; Wadsworth, 1976, ch. 7.
Thermodynamics 3, Open University Press, 1975.
General Chemistry, readings from *Scientific American*, ch. 22 (gas chromatography).

Chapter 11   *Physical chemistry*, P. W. Atkins; Oxford University Press and W. H. Freeman, 2nd edn, 1982, ch. 5.
The second law, H. A. Bent; Oxford University Press, 1965.
Basic chemical thermodynamics, E. B. Smith; Clarendon Press, Oxford, 2nd edn, 1977, ch. 3.
Advanced physics, Nuffield Course, Unit 9, Longman or Penguin, 1972.

Chapter 12   *The proton in chemistry*, R. P. Bell; Methuen, 1959.
Physical chemistry, D. Alberty and F. Daniels; Wiley, 4th edn, 1975.
General chemistry, L. C. Pauling; W. H. Freeman, 1970, chs 14 and 19.

Chapter 13   *Physical chemistry*, P. W. Atkins; Oxford University Press and W. H. Freeman, 2nd edn, 1982, chs 11 and 12.
Experimental approach to electrochemistry, N. J. Selley; Edward Arnold, 1977.
Chemistry, man and society, M. M. Jones *et al.*; W. B. Saunders, 1976.
'When your car rusts out', W. Knockemus; *J. Chem. Educ.*, **49,** 29, (1972).

Chapter 14 *Physical chemistry*, P. W. Atkins; Oxford University Press and W. H. Freeman, 2nd edn, 1982, chs 27 and 28.

Reaction kinetics, M. J. Pilling; Clarendon Press, Oxford, 1974.

General chemistry, readings from *Scientific American*, chs 15 and 16 (fast reactions, catalysis).

Heterogeneous catalysis, G. C. Bond; Clarendon Press, Oxford, 1974.

Asimov on chemistry, I. Asimov; Macdonald and Jane's, 1975, ch. 9.

Fast reactions, J. N. Bradley; Clarendon Press, Oxford, 1975.

Answers to numerical problems

Chapter

1 4 32.06
   7 3.6×10^{-19} J
      0.0072
   8 1312 kJ mol^{-1}
   9 496 kJ mol^{-1}
   11 (a) 0.27
      (b) 1.8×10^{-6}

4 3 1.95 atm
      197 kPa
   4 1.2×10^{17}
   5 0.118 Pa
   6 5.0×10^5 mol
      8.5×10^3 kg
   7 27 kg
   8 (a) 990 kPa, 990 kPa (H_2)
      (b) 1490 kPa, 990 kPa (H_2), 500 kPa (Cl_2)
      (c) 1490 kPa, 990 kPa (HCl), 500 kPa (H_2)
   9 (a) 310 m s^{-1}
      (b) 652 m s^{-1}
      0.973
   10 5850 K
   11 16 g mol^{-1}
   13 (a) 20.0 atm
      (b) 18.6 atm

5 10 LiBr 0.378
      NaBr 0.520
      KBr 0.704
      RbBr 0.760
      CsBr 0.867

6 7 15.5 mm
   8 346 g
   9 1.1×10^{23}

Chapter

7 1 0.174
      0.826
   4 (a) 2.619 kPa (CH_3OH)
          4.556 kPa (C_2H_5OH)
      (b) 7.175 kPa
      (c) 0.365 (CH_3OH)
          0.635 (C_2H_5OH)
   5 63 g
   8 60 g mol^{-1}
   9 100.05 °C
      −0.20 °C
   10 261 kPa

8 3 60 C
      3.75×10^{20}
      6.22×10^{-4} mol
   4 (a) 32.2 min
      (b) 96.5 min
   8 (a) 267 S cm^2 mol^{-1}
      (b) 145 S cm^2 mol^{-1}
      6.9 kΩ
   10 0.0188

9 3 6.6 min
   4 1.2 MJ
   5 60 g
   6 2.4 km
   9 (a) −890 kJ mol^{-1}
      (b) −3351 kJ mol^{-1}
   10 +18 kJ mol^{-1}

10 10 2.38×10^{-3} dm^6 mol^{-2}
   11 3.53×10^{-7} atm^{-2}
   12 1.89×10^{-5} mol
   13 8.36×10^{-7} atm

11 8 (a) −326.4 J K^{-1} mol^{-1}

Chapter

      (b) −199.3 J K^{-1} mol^{-1}
      (c) −477.1 J K^{-1} mol^{-1}
   9 (a) −571.6 kJ mol^{-1}
          −474.3 kJ mol^{-1}
      (b) −92.2 kJ mol^{-1}
          −32.8 kJ mol^{-1}
      (c) +922 kJ mol^{-1}
          +1064 kJ mol^{-1}
   10 (a) −474.4 kJ mol^{-1}
      (b) −33.0 kJ mol^{-1}
   11 (a) 8.0 J K^{-1} mol^{-1}
      (b) 33.1 J K^{-1} mol^{-1}

12 4 1.7×10^{-6} mol dm^{-3}
      2.0×10^{-9} mol dm^{-3}
   5 (a) 2.53
      (b) 1.42×10^{-4} mol dm^{-3}
   6 1.46

13 5 (a) +1.103 V
      (b) +0.223 V
      (c) −1.229 V
      (d) −0.152 V
   6 (a) 2×10^{37}
      (b) 3×10^7 (mol dm^{-3})^4 atm^{-1}
      (c) 8×10^{-84} atm^3 (mol dm^{-3})$^{-2}$
      (d) 3×10^{-3}

14 4 624 s
      (a) 31.2 min
      (b) 259 s
   5 (a) 0.983
      (b) 0.018
   6 13.7 min
      5.1×10^{-2} min^{-1}
   7 8.05×10^{-3} min^{-1}
      86.1 min

Index